Advanced Physics
Demystified

Demystified Series

Advanced Physics
Demystified

Stan Gibilisco

New York Chicago San Francisco Lisbon London Madrid
Mexico City Milan New Delhi San Juan Seoul
Singapore Sydney Toronto

The McGraw-Hill Companies

Library of Congress Cataloging-in-Publication Data

Gibilisco, Stan.
    Advanced physics demystified / Stan Gibilisco.—1st ed.
        p.   cm.—(Demystified series)
    Includes index.
    ISBN-13: 978-0-07-147944-8
    ISBN-10: 0-07-147944-9 (alk. paper)
    1. Physics—Popular works.   I. Title.
QC24.5.G53 2007
530—dc22

2007011393

McGraw-Hill books are available at special quantity discounts to use as premiums and sales promotions, or for use in corporate training programs. For more information, please write to the Director of Special Sales, Professional Publishing, McGraw-Hill, Two Penn Plaza, New York, NY 10121-2298. Or contact your local bookstore.

1 2 3 4 5 6 7 8 9 0   DOC/DOC   0 1 3 2 1 0 9 8 7

ISBN-13: 978-0-07-147944-8
ISBN-10: 0-07-147944-9

This book was printed on acid-free paper.

Sponsoring Editor
  Judy Bass

Editing Supervisor
  David E. Fogarty

Project Manager
  Joanna V. Pomeranz

Copy Editor
  Matthew Kushinka

Proofreader
  D & P Editorial Services, LLC

Indexer
  Stan Gibilisco

Production Supervisor
  Pamela A. Pelton

Composition
  D & P Editorial Services, LLC

Art Director, Cover
  Jeff Weeks

To Samuel, Tim, and Tony

ABOUT THE AUTHOR

Stan Gibilisco is one of McGraw-Hill's most prolific and popular authors. His clear, reader-friendly writing style makes his books accessible to a wide audience, and his experience as an electronics engineer, researcher, and mathematician makes him an ideal editor for reference books and tutorials. Stan has authored several titles for the McGraw-Hill *Demystified* library of home-schooling and self-teaching volumes, along with more than 30 other books and dozens of magazine articles. His work has been published in several languages. *Booklist* named his *McGraw-Hill Encyclopedia of Personal Computing* one of the "Best References of 1996," and named his *Encyclopedia of Electronic*s one of the "Best References of the 1980s."

CONTENTS

Contents

PREFACE

This book is for people who want to refresh or improve their knowledge of physics. The course can be used for self-teaching or as a supplement in a classroom, tutored, or home-schooling environment.

This course builds on material covered in *Physics Demystified* and is meant to serve as a continuation of that course. For that reason, I recommend that you study *Physics Demystified* before you start here. This book covers more exotic branches of physics, goes into greater depth, is more mathematical, and is written at the level of an honors physics curriculum for high-school seniors. Standard school prerequisites to this course include high-school algebra, geometry, trigonometry, first-year calculus, and first-year physics.

For those of you who want a solid mastery of math as it applies to physics, the McGraw-Hill *Demystified* Series offers several mathematics books. If you are planning to take a physics exam for college entrance, I encourage you to study as many different exam-preparation guides as you can find. You'll get some "cross-training" that way, you'll see concepts from various perspectives, and you'll be exposed to most of the notational variants that scientists and engineers use.

As you go through each chapter, you'll find practice problems with worked-out answers. Each chapter ends with a multiple-choice quiz. You may refer to the text when taking these quizzes. Because the quizzes are "open-book," some of the questions are rather difficult, but one of the choices is always "best." (I try to avoid writing trick questions.) When you're done with the quiz at the end of a particular chapter, give your list of answers to a friend. Have the friend tell you your score, but not which questions you got wrong. The answers are listed in the back of the book. Stick with a chapter until you get most, and preferably all, of the quiz answers correct.

The book concludes with a multiple-choice final exam that contains questions drawn uniformly from all the chapters. It is a "closed-book" test. Don't look back into the chapters when taking it. A satisfactory score is at least three-quarters of the answers correct. But I suggest you shoot for 90 percent! With the final exam, as with the quizzes, have a friend tell you your score without letting you know which questions you missed. That way, you won't be likely to subconsciously memorize the answers. The questions are similar in format to those you'll encounter in standardized tests.

Suggestions for future editions are welcome.

Stan Gibilisco

ACKNOWLEDGMENTS

I extend thanks to my nephew Tim Boutelle, a student at the University of Chicago. He spent many hours helping me proofread the manuscript, and he offered insights and suggestions from the point of view of the intended audience.

CHAPTER 1

Linear Motion and Plane Trajectories

Motion can be defined in terms of *vectors*, which are quantities having independent magnitude and direction. In *linear motion*, a vector can have any possible magnitude, but only the two directions defined by a specific, fixed straight line. A *plane trajectory* is a function of the position of an object in two dimensions with respect to time. It can be defined in terms of two independent *component vectors* that both lie in the same *Euclidean* (flat) plane, but not along the same straight line.

Time

Time is a dimension through which the physical world appears to change or evolve. It can be represented by itself as a *time line*, or as an axis in a coordinate system along with spatial dimensions. Time makes two events distinguishable from each other when they occur at the same point in space. In mathematical equations, time is treated as a *scalar* quantity, symbolized by the lowercase italic letter t. A scalar can vary only in terms of its size or scale. In some scientific disciplines, time can possess the property of direction such as forward/backward along a line or even orientation in multi-dimensional spaces, but we won't get into that.

The standard unit of time is the *second*, symbolized by the lowercase letter s. The second was originally defined as 1/60 of a *minute*, which is 1/60 of an *hour*, which is 1/24 of a *mean solar day*. A second was therefore considered to be 1/86,400 or 1.15741×10^{-5} of a mean solar day. That's still a fair definition, but it's problematic because the mean solar day is gradually getting longer! Nowadays the second is defined as the time required for the radiated wave corresponding to the transition between the two hyperfine levels of the ground state of the cesium-135 atom to go through 9.19263×10^9 cycles. We don't need to be concerned with the precise meaning of the jargon here, but it's worthwhile to know that the wave oscillations produced by certain atoms make excellent universal time standards.

Displacement

In linear motion, *displacement* is the difference in position or location between two points on a straight line. When displacement is defined from fixed some point P to another fixed point Q, it is a vector quantity because it has magnitude and direction. In this discussion, a general displacement vector is denoted by a lowercase boldface letter **q**. Figure 1-1 is an example of a displacement vector that lies along a line PQ in *rectangular coordinates*.

The standard unit of linear displacement magnitude is the *meter*, symbolized by the lowercase letter m. Originally, the meter was designated as the distance between two scratches on a platinum bar on display in Paris, France, representing $1/10,000,000$ (10^{-7}) of the distance from the north geographic pole to the equator as measured along the meridian running through Paris. Nowadays, the meter can be defined as the distance a beam of light travels through a perfect vacuum in 3.33564×10^{-9} seconds.

Velocity

Velocity consists of two independent components: magnitude (*speed*) and direction. Therefore, velocity is a vector quantity. The symbol for a general velocity vector

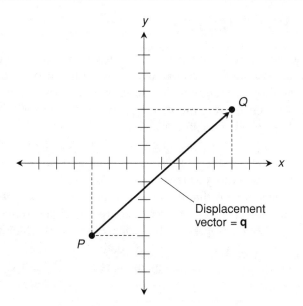

Figure 1-1   A displacement vector in a two-dimensional Cartesian coordinate system.

is a lowercase bold letter **v**. The common unit of velocity magnitude is the *meter per second* (m/s or m · s^{-1}). There is no formal absolute reference velocity because there is no absolute position or direction in the universe. However, we can consider the speed of light in a vacuum, in the absence of a gravitational field, to be a sort of "standard speed": 2.99792 × 10^8 m/s. As Albert Einstein first realized in the early 1900s, it always appears the same from any non-accelerating reference frame.

LET IT BE SO!

You will often come across statements in mathematical texts, including this book, such as: "Let **v** be a vector representing the velocity of an object *S*." This language is customary. When you are told to *let* things be a certain way, you are being asked to *suppose* that they are that way. It is an invitation to let your imagination run wild! This sets the scene in your mind for hypotheses, logical arguments, or calculations to follow.

EXPRESSIONS OF VELOCITY

Velocity can be considered as an *average* quantity over a certain period of time, or as an *instantaneous* quantity defined at a single point in time. Suppose you are driving along a straight road going due north at 25 m/s. Suddenly you see a bobcat. You hit the brakes to avoid hitting the cat, slow down to 10 m/s, watch the cat run away, and then speed up to 25 m/s again, all in a time span of one minute. Your average velocity over that minute might be 17 m/s due north. But your instantaneous velocity varies, and is 17 m/s due north at only two points in time (one as you slow down, the other as you speed back up).

Figure 1-2 shows three different velocity vectors $\mathbf{v}_1$, $\mathbf{v}_2$, and $\mathbf{v}_3$ in *rectangular three-space* for an object traveling along a straight line, speeding up as time goes by. The points t_1, t_2, and t_3 represent the instants in time at which the velocities of the object in question are $\mathbf{v}_1$, $\mathbf{v}_2$, and $\mathbf{v}_3$, respectively. The magnitude of the velocity vector in each case is represented by the length of the dashed arrowed line.

VELOCITY EQUATIONS

For constant velocity over a period of time, the following equations hold for time in seconds, displacement magnitude in meters, and velocity magnitude is in meters per second:

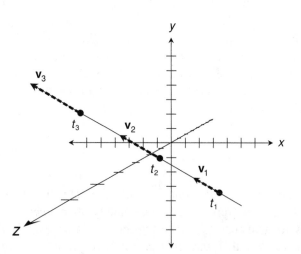

Figure 1-2 Three velocity vectors (dashed arrowed lines) in a three-dimensional rectangular coordinate system for an object traveling in a straight line.

$$\mathbf{v} = \mathbf{q} / t$$
$$\mathbf{q} = \mathbf{v}t$$
$$t = |\mathbf{q}| / |\mathbf{v}|$$

When a vector is multiplied or divided by a scalar, the direction of the vector does not change, but the magnitude is multiplied or divided by the scalar quantity. A vector symbol enclosed by vertical bars represents the magnitude of the vector. Therefore, $|\mathbf{q}|$ is the magnitude of $\mathbf{q}$, and $|\mathbf{v}|$ is the magnitude of $\mathbf{v}$. Vector magnitude is always a scalar. That's why we can get away with division in the last of the three expressions above.

Let $\mathbf{q}(t)$ be the function that defines the instantaneous displacement vector with respect to time. Let $\mathbf{v}(t)$ be the function that defines the instantaneous velocity vector with respect to time. Then:

$$\mathbf{v}(t) = d\mathbf{q}(t) / dt$$

In linear motion, the velocity and displacement vectors always lie along the same straight line, so we can simplify the above vector equation to an expression that involves only scalars. Let $|\mathbf{q}|(t)$ be the function that defines the instantaneous displacement magnitude with respect to time. Let $|\mathbf{v}|(t)$ be the function that defines the instantaneous velocity magnitude (that is, the instantaneous speed) with respect to time. Then:

$$|\mathbf{v}|(t) = d|\mathbf{q}|(t) / dt$$

We can determine $|\mathbf{q}|(t)$ on the basis of $|\mathbf{v}|(t)$ if we know the initial displacement. This formula involves an indefinite integral:

$$|\mathbf{q}|(t) = \int |\mathbf{v}|(t) \, dt$$

NEGATIVE VECTOR MAGNITUDE

In linear motion, the velocity vector always lies along the same line as the displacement vector. But it doesn't necessarily have to go in the same direction. If the point in the example of Fig. 1-2 were to stop and then back up along the same line, the velocity vector would, for awhile at least, point in exactly the opposite direction from the displacement vector. In working with physical equations, this may appear either as positive speed in the opposite direction from the displacement, or as "negative speed" in the same direction as the displacement.

This brings up a possible point of confusion. If you're traveling due north at 25 m/s and then brake, stop, shift into reverse, and back up along the same road at 5 m/s, you can express the new velocity vector in two ways. You can either say you are traveling 5 m/s due south (positive speed, opposite direction) or you can say you are traveling −5 m/s due north (negative speed, same direction).

For the purposes of our discussion, let's agree on a convention: *A vector can have negative magnitude in theory, but not in practice.* If we are doing a calculation and we come up with a vector that has theoretically negative magnitude, that's all right, but at the end of it all, let's define any vector with negative magnitude as a vector pointing in the opposite direction with a positive magnitude of the same absolute value. For example, if we derive a vector of magnitude −5 m/s pointing due north as the answer to some problem, we should call this a vector of magnitude 5 m/s pointing due south.

PROBLEM 1-1

Suppose you are thinking about buying a high-performance car. The dealer shows you a vehicle and says it can go from a dead stop to 25 m/s in exactly 5 s. You put it on a test range, hit the accelerator, take off due east, and find that it meets this standard precisely! Then you keep going for a few more seconds, but you get nervous and let up when you hit 60 m/s. Now suppose that during the first 5-s period, the magnitude of the velocity vector increases at a constant rate. Also suppose that you drive in a straight line on level ground. What is the instantaneous velocity of the car 3 s after you first hit the accelerator from a dead stop?

SOLUTION 1-1

In order to solve this, you must derive a formula that describes the velocity as a function of time. Because the car moves in a straight line, all you need to do is find a speed-vs.-time function. You know that all the vectors point due east. After a little cogitation, you come up with this:

$$|\mathbf{v}| = 5t$$

This is a simple example of a *linear equation*. The speed changes at a constant rate. You know that 5 is the correct *coefficient* for this linear equation because, when $t = 5$, $|\mathbf{v}| = 25$ m/s. (That's what the dealer told you, and you confirmed it during your driving experiment!) At $t = 3$ s, therefore, $|\mathbf{v}| = 5 \times 3 = 15$ m/s, and therefore $\mathbf{v} = 15$ m/s due east.

Acceleration

Acceleration is an expression of the change in the velocity of an object. Therefore, acceleration can manifest as a change in speed, a change in direction, or both. In linear motion, the acceleration vector always lies along the same line as the velocity and displacement vectors. When speed increases in linear motion, the acceleration vector points in the same direction as the velocity vector. The term *deceleration* is sometimes used loosely by lay people (but not often by physicists!) to describe the motion of an object traveling in a straight line while losing speed. In this case, the acceleration vector points in a direction exactly opposite that of the velocity vector.

EXPRESSIONS OF ACCELERATION

Acceleration is a vector quantity consisting of two independent components: magnitude (rate of change in speed) and direction. The symbol for a general acceleration vector is a lowercase bold letter **a**. The common unit of acceleration magnitude is the *meter per second squared* (m/s^2 or $m \cdot s^{-2}$). Acceleration, like velocity, can be averaged over a period of time, or expressed as an instantaneous quantity at a single point in time.

ACCELERATION EQUATIONS

For constant acceleration over a period of time, the following equations hold for time in seconds, velocity in meters per second, and acceleration in meters per second squared:

$$\mathbf{a} = \mathbf{v} / t$$

$$\mathbf{v} = \mathbf{a}t$$

$$t = |\mathbf{v}| / |\mathbf{a}|$$

Let $\mathbf{a}(t)$ be the function that defines the instantaneous acceleration vector with respect to time. Then:

$$\mathbf{a}(t) = d\mathbf{v}(t) / dt$$

In linear motion, the acceleration and velocity vectors always lie along the same straight line, so we can simplify the above vector equation to an expression that involves only scalars. Let $|\mathbf{a}|(t)$ be the function that defines the instantaneous acceleration magnitude with respect to time. Then:

$$|\mathbf{a}|(t) = d|\mathbf{v}|(t) / dt$$

We can determine $|\mathbf{v}|(t)$ on the basis of $|\mathbf{a}|(t)$ if we know the initial velocity. Again, this formula involves an indefinite integral:

$$|\mathbf{v}|(t) = \int |\mathbf{a}|(t) \, dt$$

It's possible to derive $|\mathbf{q}|(t)$ on the basis of $|\mathbf{a}|(t)$ if we know the initial velocity and displacement. We integrate the acceleration function twice with respect to time:

$$|\mathbf{q}|(t) = \iint |\mathbf{a}|(t) \, dt \, dt$$

The acceleration function is integrated to obtain the velocity function, which is in turn integrated to obtain the displacement function. The initial velocity and the initial displacement determine the values of the constants of integration.

PROBLEM 1-2

Imagine that you are in the gondola of a balloon hovering in a fixed spot 1.000 km above a deserted lake. There is no wind. You hold a brick over the side and let it go. Suppose the brick is so dense that air resistance does not affect it before splashdown. (In real life, an object falling in the atmosphere reaches a certain maximum *terminal velocity* because of friction with the air. Once the object reaches terminal velocity in free fall, it cannot gain any more speed because the force of atmospheric friction balances the force caused by gravitation.) For simplicity, consider the *acceleration of gravity* vector to have a magnitude of exactly 10.00 m/s². Plot a graph showing the instantaneous downward acceleration magnitude and the instantaneous downward velocity magnitude for the first 10.00 s of the brick's descent. Then use integration to derive the functions of instantaneous downward velocity magnitude vs. time and instantaneous downward displacement magnitude vs. time. How long will it take for the brick to splash down after it is dropped?

SOLUTION 1-2

The solid line in Fig. 1-3 shows the instantaneous downward acceleration magnitude function, $|\mathbf{a}|(t)$. It is a constant 10.00 m/s². The dashed line in Fig. 1-3 shows the instantaneous downward velocity magnitude function, $|\mathbf{v}|(t)$. The initial velocity is zero, and it increases downward by 10.00 m/s every second, indefinitely. (That is what 10.00 m/s² means: 10.00 meters per second, per second.) From the graph we see that:

$$|\mathbf{a}|(t) = 10.00$$
$$|\mathbf{v}|(t) = 10.00t$$

Rather than examining the graph and intuitively plotting $|\mathbf{v}|(t)$, you can derive $|\mathbf{v}|(t)$ by integrating $|\mathbf{a}|(t)$ over time:

$$|\mathbf{v}|(t) = \int |\mathbf{a}|(t) \, dt$$
$$= \int 10.00 \, dt$$
$$= 10.00t + c$$

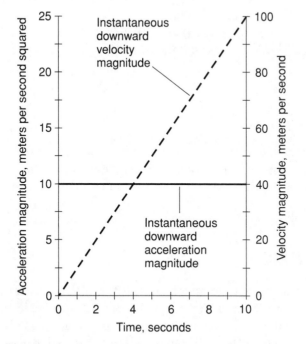

Figure 1-3   Illustration for Problem and Solution 1-2.

The value of c, the *constant of integration*, is zero in this case because the initial velocity is zero. Therefore:

$$|\mathbf{v}|(t) = 10.00t$$

You can find the function $|\mathbf{q}|(t)$ that describes the distance the brick has fallen after a given period of time by integrating the function $|\mathbf{v}|(t)$ over time, setting the constant of integration (which represents the initial position) equal to zero:

$$\begin{aligned}
|\mathbf{q}|(t) &= \int |\mathbf{v}|(t)\ dt \\
&= \int 10.00t\ dt \\
&= 5.000t^2 + c \\
&= 5.000t^2
\end{aligned}$$

You want to know how long it will take the brick to hit the water after you release it. This is the time t at which $|\mathbf{q}|(t) = 1.000 \times 10^3$, the altitude of the balloon in meters. Therefore:

$$5.000t^2 = 1.000 \times 10^3$$
$$t^2 = 1.000 \times 10^3 / 5.000 = 200.0$$
$$t = 200.0^{1/2} = 14.14 \text{ s}$$

Mass

Mass is an expression of the amount of matter in an object or region of space. The most basic unit of mass is the *atomic mass unit* (amu or u), which is exactly 1/12 the mass of a carbon-12 (C-12) nucleus. All other units of mass can be derived from this.

UNITS OF MASS

Physicists express mass in terms of *grams* (g) or *kilograms* (kg), where 1 g = 0.001 kg and 1 kg = 1000 g. A mass of 1 g is 6.02214×10^{23} amu, rounded to six significant figures. This is the *Avogadro constant*, also called a *mole*. A mass of 1 kg is 1000 g or 6.02214×10^{26} amu. This is called a *kilomole*. A gram is equal to a mole of atomic mass units, and a kilogram is equal to a kilomole of atomic mass units. The following equation expresses the atomic mass unit in terms of the gram and the kilogram, to six significant figures according to the *National Institute of Standards and Technology* (NIST):

$$1 \text{ amu} = 1.66054 \times 10^{-24} \text{ g}$$
$$= 1.66054 \times 10^{-27} \text{ kg}$$

Small masses are sometimes expressed in *milligrams* (mg), *micrograms*, (µg), *nanograms* (ng), or *picograms* (pg), where:

$$1 \text{ mg} = 0.001 \text{ g}$$
$$1 \text{ µg} = 0.001 \text{ mg} = 10^{-6} \text{ g}$$
$$1 \text{ ng} = 0.001 \text{ µg} = 10^{-9} \text{ g}$$
$$1 \text{ pg} = 0.001 \text{ ng} = 10^{-12} \text{ g}$$

Large masses can be expressed in terms of *megagrams* (Mg), *gigagrams* (Gg), or *teragrams* (Tg), where:

$$1 \text{ Mg} = 1000 \text{ kg} = 10^6 \text{ g}$$
$$1 \text{ Gg} = 1000 \text{ Mg} = 10^9 \text{ g}$$
$$1 \text{ Tg} = 1000 \text{ Gg} = 10^{12} \text{ g}$$

PREFIX MULTIPLIERS

Are some of the above mentioned units alien to you? That's all right; the larger ones are rarely used. It is common practice among physicists to state all mass values in grams or kilograms, and to use scientific notation for values less than 0.001 or larger than 1000. Nevertheless, you will occasionally come across *prefix multipliers* for units of various quantities. Table 1-1 lists the power-of-10 prefix multipliers from 10^{-24} to 10^{24}. (There's another set of prefix multipliers based on powers of 2. They are used in computer applications.)

MASS IS A SCALAR

In equations and formulas, mass has magnitude but not direction. Although mass itself is a scalar, quantities or phenomena with mass as a component can be vectors

Table 1-1   Prefix multipliers used in physical sciences and engineering

Prefix	Symbol	Multiplier
yocto-	y	10^{-24}
zepto-	z	10^{-21}
atto-	a	10^{-18}
femto-	f	10^{-15}
pico-	p	10^{-12}
nano-	n	10^{-9}
micro-	μ or mm	10^{-6}
milli-	m	0.001
centi-	c	0.01
kilo-	k	1000
mega-	M	10^{6}
giga-	G	10^{9}
tera-	T	10^{12}
peta-	P	10^{15}
exa-	E	10^{18}
zetta-	Z	10^{21}
yotta-	Y	10^{24}

if one of the other components is a vector. A good example is *weight*, which is the product of mass (a scalar) and the acceleration of gravity (a vector).

PROBLEM 1-3
Imagine a particle of dust that masses 25.48 μg. What is its mass in atomic mass units? Round off the answer to the appropriate number of significant figures.

SOLUTION 1-3
Note that 1 g = 6.02214×10^{23} amu, and 1 μg = 10^{-6} g. From these facts, it is evident that 1 μg = 6.02214×10^{17} amu. (Subtract 6 from the power of 10 in scientific notation.) Therefore:

$$25.48 \ \mu g = 25.48 \times 6.02214 \times 10^{17} \text{ amu}$$
$$= 153.4 \times 10^{17} \text{ amu}$$
$$= 1.534 \times 10^{19} \text{ amu}$$

This is rounded to four significant figures because that is the number of significant figures in the least precise numerical input data.

Force, Momentum, and Impulse

In physics, the term *force* refers to any phenomenon that can change the velocity of a mass. A region in which a single force acts is called a *force field*. Force is a vector quantity. The symbol for a general force vector is the uppercase bold letter **F**.

UNITS OF FORCE

The standard unit of force magnitude is the *newton* (N). By definition, 1 N is the force that causes a mass of exactly 1 kg to accelerate at a rate of exactly 1 m/s^2. A newton is equivalent to a kilogram meter per second squared (kg · m/s^2 or kg · m · s^{-1}). Small forces can be expressed in *millinewtons* (mN), *micronewtons*, (μN), *nanonewtons* (nN), or *piconewtons* (pN), where:

$$1 \text{ mN} = 0.001 \text{ N}$$
$$1 \ \mu N = 0.001 \text{ mN} = 10^{-6} \text{ N}$$
$$1 \text{ nN} = 0.001 \ \mu N = 10^{-9} \text{ N}$$
$$1 \text{ pN} = 0.001 \text{ nN} = 10^{-12} \text{ N}$$

Large forces can be expressed in terms of *kilonewtons* (kN), *meganewtons* (MN), *giganewtons* (GN), or *teranewtons* (TN), where:

$$1 \text{ kN} = 1000 \text{ N}$$
$$1 \text{ MN} = 1000 \text{ kN} = 10^6 \text{ N}$$
$$1 \text{ GN} = 1000 \text{ MN} = 10^9 \text{ N}$$
$$1 \text{ TN} = 1000 \text{ GN} = 10^{12} \text{ N}$$

FORMS OF FORCE

When most people think of force, they imagine *mechanical force* such as a woman pushing a door open or the earth's gravitation pulling downward on a man's body. However, force fields can also be produced by other phenomena such as electrostatic charges, magnetic poles, the interaction among particles in atomic nuclei, and high-speed particle bombardment. It is reasonable to suppose that force fields exist in forms that humans haven't yet directly observed or quantified.

Force can be expressed mathematically as the product of mass and acceleration. The vector equation is:

$$\mathbf{F} = m\mathbf{a}$$

for mass in kilograms, acceleration magnitude in meters per second squared, and force magnitude in newtons. In linear motion, we can state this formula in terms of force and acceleration magnitude only, so the equation contains only scalar quantities:

$$|\mathbf{F}| = m|\mathbf{a}|$$

MASS, ACCELERATION, AND WEIGHT

Mass becomes *weight* in the presence of acceleration or gravitation. Weight is a mechanical force, is properly expressed in newtons, and is a vector quantity. In any situation involving a mass that exhibits weight, the direction of the weight vector is the same as the direction of the force vector.

On the earth's surface, the weight of an object is sometimes specified in kilograms, and is taken to be equal to the mass of that object in kilograms. But that is an oversimplification. The "weight in kilograms" notion won't work on any planet where the gravitational field at the surface is more or less intense than that on earth. In fact, the weight of a given mass varies slightly even with changes in location on our planet, because the gravitational field is a little bit stronger at some points on the earth's surface than at others.

In the United States, weight is usually expressed in units called *pounds*, where a pound is equal to approximately 0.454 kg. That's good enough for a bathroom scale. An instrument calibrated in pounds can tell you if you're getting more or less massive, but it's technically imprecise. A true weight scale would be calibrated in newtons, not pounds or kilograms.

MOMENTUM

Momentum is defined as the product of an object's mass and its velocity. Because velocity is a vector, momentum is a vector too. Momentum magnitude is expressed in kilogram meters per second (kg · m/s or kg · m · s^{-1}). If we let **v** represent the velocity vector and **p** represent the momentum vector, then:

$$\mathbf{p} = m\mathbf{v}$$

for mass in kilograms, velocity magnitude in meters per second, and momentum magnitude in kilogram meters per second. Linear motion involves magnitudes only:

$$|\mathbf{p}| = m|\mathbf{v}|$$

Let $\mathbf{F}(t)$ define the instantaneous force vector as a function of time, and let $\mathbf{p}(t)$ define the instantaneous momentum vector as a function of time. Then the instantaneous value of $\mathbf{F}(t)$ is equal to the instantaneous rate of change in $\mathbf{p}(t)$ with respect to time:

$$\mathbf{F}(t) = d\mathbf{p}(t) \,/\, dt$$

In linear motion, the vectors all point along a single straight line, so we can deal with scalar quantities. Let $|\mathbf{F}|(t)$ define the instantaneous force magnitude as a function of time. Let $|\mathbf{p}|(t)$ define the instantaneous momentum magnitude as a function of time. Then:

$$|\mathbf{F}|(t) = d|\mathbf{p}|(t) \,/\, dt$$

The linear momentum vector of an object is constant if and only if the net force vector acting on that object is zero. This is the *law of conservation of linear momentum*. When the net force vector on a solid object (the vector sum of all the forces, if there are more than one) is zero, that object is said to be in *translational equilibrium*.

IMPULSE

Impulse, symbolized by the uppercase bold letter **J**, consists of force acting for a period of time, causing a change in the momentum of a mass. If the force remains constant, the vector and scalar equations defining impulse are:

$$\mathbf{J} = \mathbf{F}t$$
$$|\mathbf{J}| = |\mathbf{F}|t$$

for force magnitude in newtons, time in seconds, and impulse magnitude in kilogram meters per second (kg · m/s or kg · m · s^{-1}). If we substitute $m\mathbf{a}$ for **F** in the equation for impulse and rearrange the order of the variables, we get this vector-scalar product:

$$\mathbf{J} = (mt)\mathbf{a}$$

In linear motion, we can substitute the vector magnitudes for the vector quantities, obtaining this all-scalar product:

$$|\mathbf{J}| = mt|\mathbf{a}|$$

Have you noticed that impulse and momentum are both defined in kilogram meters per second (kg · m/s or kg · m · s^{-1})? This doesn't mean that the two phenomena are identical, but they are related. When a certain impulse is imparted to a moving mass, the momentum of that mass changes.

In some situations, the force applied to a moving object changes with time. In scenarios like this, we need to resort to integration in order to calculate impulse. Let $|\mathbf{J}|(t)$ define the accumulated impulse magnitude as a function of time. Let $|\mathbf{F}|(t)$ define the instantaneous force magnitude as a function of time. Then $|\mathbf{J}|(t)$ is equal to the integral of $|\mathbf{F}|(t)$ with respect to time:

$$|\mathbf{J}|(t) = \int |\mathbf{F}|(t)\, dt$$

NEWTON'S FIRST LAW OF MOTION

In the 1600s, the English scientist and mathematician *Isaac Newton* formulated three rigorous rules that govern the way all things move (as long as relativistic effects are not involved). *Newton's first law of motion*, sometimes called *Galileo's law*, can be broken down into two statements:

- A mass at rest with respect to a non-accelerating observer remains at rest relative to that observer unless acted upon by an external force
- A mass in motion with respect to a non-accelerating observer maintains a constant velocity relative to that observer unless acted upon by an external force

NEWTON'S SECOND LAW OF MOTION

Newton's second law of motion can be stated as follows. Suppose a mass of *m* kilograms is acted on by a constant force of magnitude $|\mathbf{F}|$ in newtons, causing a constant, straight-line acceleration of magnitude $|\mathbf{a}|$ in meters per second squared. Then the force, mass, and acceleration are related according to the following formula:

$$|\mathbf{F}| = m\,|\mathbf{a}|$$

NEWTON'S THIRD LAW OF MOTION

Newton's third law of motion has been heard and memorized by almost everyone: For every action, there is an equal and opposite reaction. If $\mathbf{F}$ represents a force vector acting on a given body, then the reaction force vector $\mathbf{G}$ is related to $\mathbf{F}$ as follows:

$$\mathbf{G} = -\mathbf{F} = (-1)\mathbf{F}$$

In scalar terms:

$$|\mathbf{G}| = |\mathbf{F}|$$

This reflects the fact that although the action and reaction vectors directly oppose each other in direction, they have the same magnitude.

PROBLEM 1-4
Imagine that you are on an interplanetary vessel coasting through space to a rendezvous with Neptune. You discover that an immediate course correction is necessary. Otherwise the ship will not fall into orbit around Neptune, but will fly off into interstellar space! Your onboard computers indicate that the ship must gain 100.0 m/s of forward speed to get back on course. Suppose the mass of the craft is exactly 2.000×10^4 kg. The rockets produce a constant force of 2.500×10^4 N. When the rockets are fired, how fast will the ship accelerate? How long will it take for the forward velocity of the ship to increase by the required 100.0 m/s? What will be the impulse that produces this course correction?

SOLUTION 1-4
In order to determine the rate of acceleration, use Newton's second law:

$$|\mathbf{F}| = m\,|\mathbf{a}|$$
$$|\mathbf{a}| = |\mathbf{F}|\,/\,m$$
$$|\mathbf{a}| = (2.500 \times 10^4)\,/\,(2.000 \times 10^4)$$
$$= 1.250 \text{ m/s}^2$$

The ship accelerates at a constant rate of 1.250 m/s^2. It will take 80.00 s for its velocity to increase by 100.0 m/s because:

$$1.250 \text{ m/s}^2 \times 80.00 \text{ s} = 100.0 \text{ m/s}$$

The impulse magnitude, $|\mathbf{J}|$, is equal to the product of the force magnitude and the time:

$$
\begin{aligned}
|\mathbf{J}| &= |\mathbf{F}|\, t \\
&= 2.500 \times 10^4 \times 80.00 \\
&= 2.000 \times 10^6 \text{ kg} \cdot \text{m/s}
\end{aligned}
$$

All of the vectors point in the same direction, that is, along the line corresponding to the path of the vessel through space in the short-term sense. (As the ship moves outward through the solar system on its way from the earth to Neptune, its actual path is a long arc. But during the few seconds of the course correction process, it can be considered as a straight line.)

Plane Trajectories

When an object follows a trajectory in a Euclidean plane, its motion can be described in terms of linear-motion vectors. In the coordinate xy-plane, the component vectors can be defined along the x-axis and the y-axis. No matter how complicated the displacement, velocity, and acceleration functions are, they can always be broken down into independent *x-component* and *y-component* linear displacement, velocity, and acceleration functions.

COMPONENT VECTORS IN TWO DIMENSIONS

Imagine an object that follows a curved path starting at the point $(x,y) = (0,0)$, or the *origin*, in the xy-plane. Let's call the object and its location point S. Figure 1-4 shows an example. The displacement vector $\mathbf{q}$ points radially outward from the origin. When its *originating point* (or *back-end point*) coincides with the origin of the coordinate system, $\mathbf{q}$ can be defined as the ordered pair (x,y) representing the coordinate values of its *terminating point* (or *end point*).

The displacement vector $\mathbf{q}$ can be broken down into two components. One component, $\mathbf{q}_x$, lies on the x axis. The other component, $\mathbf{q}_y$, lies on the y axis. The originating points of both components are $(0,0)$. The two components add up vectorially to $\mathbf{q}$, as follows:

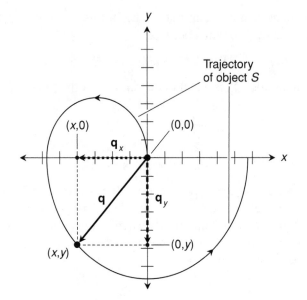

Figure 1-4   The displacement vector **q**, and its component vectors $\mathbf{q}_x$ and $\mathbf{q}_y$, at a point S in the trajectory of an object traveling in the xy-plane.

$$\mathbf{q} = \mathbf{q}_x + \mathbf{q}_y$$

Because the originating points of $\mathbf{q}_x$ and $\mathbf{q}_y$ both coincide with the coordinate origin, we can define the displacement components, and their composite, according to their terminating points:

$$\mathbf{q}_x = (x,0)$$
$$\mathbf{q}_y = (0,y)$$
$$\mathbf{q} = (x,y) = (x,0) + (0,y)$$

Now consider the displacement function $\mathbf{q}(t)$ of S with respect to time. This is a function in which two variables, x and y, change independently. Therefore, $\mathbf{q}(t)$ can be broken down into two different scalar functions of time. Let $\mathbf{q}_x(t)$ be the function representing the value of x with respect to time, and let $\mathbf{q}_y(t)$ be the function representing the value of y with respect to time. Then:

$$\mathbf{q}(t) = \mathbf{q}_x(t) + \mathbf{q}_y(t)$$

The example in Fig. 1-4 shows displacement vectors and components. The same principles apply to velocity and acceleration vectors and components. In general:

- Any two-dimensional motion vector or function can be broken down into two separate and independent component vectors or functions
- Separate calculations and operations can be carried out for the two component vectors or functions, treating them as independent cases of linear motion
- When all the calculating is done, the two-dimensional solution can be stated as the vector sum of the two linear solutions

MAGNITUDE AND DIRECTION IN TWO DIMENSIONS

A vector in the *xy*-plane can be defined according to the coordinates of its terminating point, as we have done in the example of Fig. 1-4. However, sometimes we want to define a two-dimensional vector in terms of its magnitude and direction. Then we must use the two-dimensional *distance formula* to figure out the magnitude in linear units, and trigonometry to express the direction as an angle.

Figure 1-5 shows the same vector displacement vector **q** as described above, but in terms of its magnitude and direction rather than in terms of its components. The magnitude of **q**, written |**q**|, is given by the following formula:

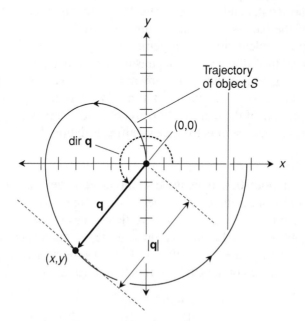

Figure 1-5   Magnitude and direction of the displacement vector **q** at a point *S* in the trajectory of an object traveling in the *xy*-plane.

$$|\mathbf{q}| = (x^2 + y^2)^{1/2}$$

where the 1/2 power represents the positive square root.

When x is positive, the direction of $\mathbf{q}$ (written dir $\mathbf{q}$) is the angle that $\mathbf{q}$ subtends with respect to the positive x axis, expressed in degrees counterclockwise as follows:

$$\text{dir } \mathbf{q} = \arctan (y/x)$$

When x is negative, the formula for the direction of $\mathbf{q}$ is:

$$\text{dir } \mathbf{q} = 180° + \arctan (y/x)$$

If $x = 0$, then dir $\mathbf{q} = 90°$ when y is positive and $270°$ when y is negative. If $x = 0$ and $y = 0$, then $\mathbf{q}$ is the *zero vector*, and dir $\mathbf{q}$ is undefined.

PROBLEM 1-5

Suppose a baseball player stands on a huge, level field and throws a ball at a 45° angle relative to the horizon at an initial speed of 70.711 m/s. Let the x axis of a coordinate plane represent the horizontal component of the ball's path. Let the y axis represent the vertical component. Define the starting point of the trajectory, where the player stands and hurls the ball, as the coordinate origin (0,0), disregarding the height of the player's hand above the ground when he releases the ball. What are the horizontal and upward-vertical velocity and acceleration magnitude components of the ball as functions of time, neglecting the effects of air resistance? To keep the calculations from getting messy, consider the acceleration of gravity to be 10.0 m/s^2 downward, which is theoretically equivalent to -10.0 m/s^2 upward. Draw a graph showing the upward-vertical velocity and acceleration magnitude components as functions of time. What is the upward-vertical displacement magnitude of the ball as a function of time? Draw a graph showing this function. Express your answers to three significant figures.

SOLUTION 1-5

The horizontal acceleration magnitude component is zero at all times, so the horizontal velocity magnitude component is constant. (The reason for this is explained in Solution 1-7.) The upward-vertical acceleration magnitude component is a constant -10 m/s. Let $|\mathbf{a}_x|(t)$ represent the horizontal acceleration magnitude component as a function of time. Let $|\mathbf{a}_y|(t)$ represent the upward-vertical acceleration magnitude component as a function of time. Then:

$$|\mathbf{a}_x|(t) = 0.00$$
$$|\mathbf{a}_y|(t) = -10.0$$

The latter of these facts is illustrated in Fig. 1-6 as a solid horizontal line.

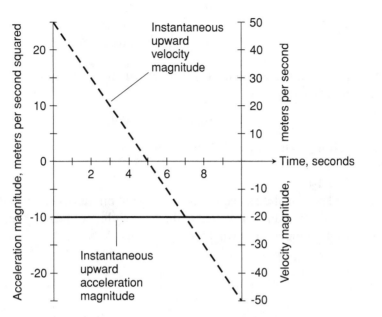

Figure 1-6   Illustration for Problem and Solution 1-5.

Because the ball is thrown at a 45° angle, the initial horizontal velocity magnitude component, $|\mathbf{v}_x|$, is the same as the initial upward-vertical velocity magnitude component $|\mathbf{v}_y|$. Working the formula for the magnitude of a vector backwards:

$$|\mathbf{v}_x|^2 + |\mathbf{v}_y|^2 = 70.711^2 = 5000$$
$$|\mathbf{v}_x| = 50.0 \text{ m/s}$$
$$|\mathbf{v}_y| = 50.0 \text{ m/s}$$

As time passes, $|\mathbf{v}_x|$ is always 50.0 m/s, but $|\mathbf{v}_y|$ changes at a constant rate of -10.0 m/s^2. If we let $|\mathbf{v}_x|(t)$ represent the horizontal velocity magnitude component as a function of time and $|\mathbf{v}_y|(t)$ represent the upward-vertical velocity magnitude component as a function of time, then:

$$|\mathbf{v}_x|(t) = 50.0$$
$$|\mathbf{v}_y|(t) = 50.0 - 10.0t$$

The latter of these facts is illustrated in Fig. 1-6 as a dashed slanted line.

The upward-vertical displacement magnitude as a function of time, $|\mathbf{q}_y|(t)$, is obtained by integrating $|\mathbf{v}_y|(t)$ with respect to t, as follows:

$$|\mathbf{q}_y|(t) = \int |\mathbf{v}_y|(t) \, dt$$
$$= \int 50.0 - 10.0t \, dt$$
$$= 50.0t - 5.00t^2 + c$$

Because the starting displacement is at the origin, the constant of integration, c, is equal to zero. Therefore:

$$|\mathbf{q}_y|(t) = 50.0t - 5.00t^2$$

Figure 1-7 is a graph showing this function.

PROBLEM 1-6

What is the horizontal distance between the point on the ground where the ball is first thrown (disregarding the height of the player's hand when he releases the ball) and the point on the ground where the ball lands? What is the maximum altitude

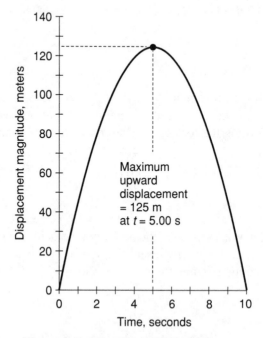

Maximum
upward
displacement
= 125 m
at $t = 5.00$ s

Figure 1-7 Illustration for Problems and Solutions 1-5 and 1-6.

attained by the ball? Draw the path of the ball as it travels from the origin to point S where it hits the ground. Draw the displacement vector **q** at the midpoint M of the ball's trajectory. Calculate its direction and magnitude at point M. Draw approximations of the velocity and acceleration vectors **v** and **a**, and calculate their magnitudes and directions at point M. Express your answers to three significant figures.

SOLUTION 1-6
Figure 1-7 tells us that the ball is aloft for 10.0 s. We can verify that this is true by plugging the value $t = 10.0$ into the function of upward-vertical displacement magnitude vs. time and making sure the result is equal to zero:

$$|\mathbf{q}_y|(t) = 50.0t - 5.00t^2$$
$$|\mathbf{q}_y|(10.0) = 50.0 \times 10.0 - 5.00 \times (10.0)^2$$
$$= 500 - 500$$
$$= 0.00 \text{ m}$$

The horizontal velocity magnitude component $|\mathbf{v}_x|$ stays constant, and we found that it is initially equal to 50.0 m/s. Therefore, the ball stays aloft for a total distance of 50.0 m/s × 10.0 s = 500 m. A major-league baseball franchise owner would pay a lot of money for a center fielder who could throw a ball that far! The ball reaches its highest altitude at the midpoint of its flight, or $t = 5.00$ s, because $t = 5.00$ represents the local maximum of $|\mathbf{q}_y|(t)$. If you'd like an "extra credit" exercise, prove this using the technique for finding local maxima and minima that you learned in first-year calculus. Plugging $t = 5.00$ into the vertical displacement magnitude function gives us the following result:

$$|\mathbf{q}_y|(t) = 50.0t - 5.00t^2$$
$$= 50.0 \times 5.00 - 5.00 \times 5.00^2$$
$$= 250 - 125$$
$$= 125 \text{ m}$$

Figure 1-8 is a drawing, to scale in the xy-plane, of the arc followed by the ball from the point (0,0) where it is thrown to the point S where it lands. The displacement vector **q** at the midpoint M is shown. Its terminating point is $(x,y) = (250,125)$. Using the direction formula to calculate dir **q**, we obtain:

$$\text{dir } \mathbf{q} = \arctan (y/x)$$
$$= \arctan (125/250)$$
$$= \arctan 0.500$$
$$= 26.6°$$

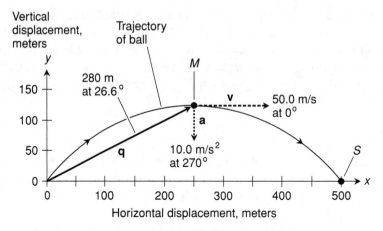

Figure 1-8   Illustration for Problem and Solution 1-6.

Using the formula for vector magnitude based on its terminating point (x,y) when the originating point is at the origin, we get this:

$$|\mathbf{q}| = (x^2 + y^2)^{1/2}$$
$$= (250^2 + 125^2)^{1/2}$$
$$= (62{,}500 + 15{,}625)^{1/2}$$
$$= 78{,}125^{1/2}$$
$$= 280 \text{ m}$$

At point M, the horizontal velocity magnitude component $|\mathbf{v}_x|$ is 50.0 m/s, as it is at all times. Its instantaneous upward-vertical velocity component at point M, where $t = 5.00$ s, can be determined from the function $|\mathbf{v}_y|(t)$ we derived earlier:

$$|\mathbf{v}_y|(t) = 50.0 - 10.0t$$
$$= 50.0 - 10.0 \times 5.00$$
$$= 50.0 - 50.0$$
$$= 0.00 \text{ m/s}$$

Therefore, at point M, the velocity vector $\mathbf{v}$ points along the positive x axis with a magnitude of 50.0 m/s:

$$\text{dir } \mathbf{v} = 0.00°$$
$$|\mathbf{v}| = 50.0 \text{ m/s}$$

From Solution 1-5, we know these two facts about the acceleration magnitude components as functions of time:

$$|\mathbf{a}_x|(t) = 0.00$$
$$|\mathbf{a}_y|(t) = -10.0$$

These are both constant functions. At M, as at all other points along the trajectory, the acceleration vector $\mathbf{a}$ points straight down at an angle of 270° and has a magnitude of 10.0 m/s². (Remember the discussion about negative vector magnitudes. A theoretical value of –10.0 units upward is equivalent in the real world to 10.0 units downward.) Therefore:

$$\text{dir } \mathbf{a} = 270°$$
$$|\mathbf{a}| = 10.0 \text{ m/s}^2$$

Figure 1-8 illustrates this situation. The midpoint of the trajectory is M. The point of impact is S. The trajectory and vector $\mathbf{q}$ are drawn to scale. Vectors $\mathbf{v}$ and $\mathbf{a}$ are portrayed in their true directions, but their magnitudes are exaggerated and their originating points have been moved to M for clarity.

PROBLEM 1-7

A claim is made in both of the previous two solutions that the ball's horizontal velocity magnitude component is constant at 50.0 m/s. Much of what we derive later is based on this claim. How do we know it is true?

SOLUTION 1-7

When the ball is first thrown, the horizontal velocity magnitude component is 50.0 m/s, as we derived in Solution 1-5. The magnitude $|\mathbf{F}_x|$ of the horizontal force acting on the ball is zero after the player lets it go. (Gravity constantly pulls vertically on the ball, but nothing pushes it or pulls on it horizontally. Remember that we're neglecting the effects of air resistance!) Because there is no horizontal force component, Newton's first law of motion tells us that the horizontal velocity magnitude component cannot change. It starts out at 50.0 m/s, so it must remain at 50.0 m/s as long as the ball is aloft.

Because there is vertical acceleration, the vertical component of the velocity magnitude is not constant. The speed of the ball along its actual arc, called the *tangential speed*, is the vector sum of the horizontal and vertical velocity magnitude components. For this reason, the absolute speed of the ball does, in fact, change with time. It is greatest at the beginning and end of the trajectory, and is smallest at the midpoint. If you'd like another "extra credit" exercise, derive a function that describes the tangential speed of the ball with respect to time!

Quiz

This is an "open book" quiz. You may refer to the text in this chapter. A good score is 8 correct. Answers are in the back of the book.

1. Suppose you are driving a car on a level, straight road due east at 50 kilometers per hour (km/h). You hit the brakes, slowing down to 30 km/h in a few seconds. During this time, in theory, your acceleration vector magnitude can be considered

   (a) positive, in an eastward direction.

   (b) negative, in an eastward direction.

   (c) negative, in a westward direction.

   (d) zero.

2. Imagine that you toss a ball straight up at a speed of 20 m/s. Consider the acceleration of gravity to be exactly 10 m/s^2. How long will it take for the ball to fall back to the ground, neglecting air resistance or wind effects?

   (a) 2.0 s.

   (b) 3.0 s.

   (c) 4.0 s.

   (d) 6.0 s.

3. Consider again the ball in the previous problem. Which of the following statements (a), (b), or (c), if any, is false?

   (a) Between the time the ball is released and the time it returns to the ground, it can be said to decelerate for half the time and accelerate for half the time.

   (b) Between the time the ball is released and the time it returns to the ground, it accelerates downward at a constant rate.

   (c) Between the time the ball is released and the time it returns to the ground, its velocity constantly changes.

   (d) All of the above statements (a), (b), and (c) are true.

4. Imagine an object that starts moving along a straight line from a dead stop. The instantaneous displacement magnitude (in meters) varies according to twice the cube of the elapsed time (in seconds) from the instant it starts moving. How fast is the object moving after 2.00 s?

(a) 8.00 m/s.

(b) 16.0 m/s.

(c) 24.0 m/s.

(d) More information is needed to answer this.

5. Suppose an object is traveling at an initial velocity of 100 m/s in a straight line. It decelerates (slows down) at a constant rate of 500 m/s². How long will it take for the instantaneous speed to drop to zero?

(a) 200 ms.

(b) 2.00 s.

(c) 5.00 s.

(d) More information is needed to answer this.

6. Suppose you hurl a stone horizontally outward from the base of a cliff that overlooks the ocean from half a kilometer (0.50 km) up. Neglecting the effects of air resistance, how long after you release the stone will it splash down?

(a) 10 s.

(b) 20 s.

(c) 50 s.

(d) It depends on the initial horizontal velocity magnitude component of the stone.

7. Imagine a linear-motion unit called the meter per second cubed (m/s³ or m · s⁻³). This represents

(a) the rate of change in displacement magnitude with respect to time.

(b) the rate of change in velocity magnitude with respect to time.

(c) the rate of change in acceleration magnitude with respect to time.

(d) Forget it! Such a unit cannot represent any phenomenon in linear motion.

8. Consider an object whose initial displacement and velocity are both equal to zero with respect to a reference point X. The object accelerates in a straight line away from X according to the following function of time:

$$|\mathbf{a}|(t) = 2.000t$$

where the instantaneous acceleration is expressed in meters per second squared, and the initial time is $t = 0$. How fast is this object moving with respect to point X at $t = 3.000$ s?

(a) 1.732 m/s.

(b) 3.000 m/s.

(c) 6.000 m/s.

(d) 9.000 m/s.

9. Consider again the object whose acceleration is described in Question 8. How far is it from point X at $t = 3.000$ s?

(a) 3.000 m.

(b) 9.000 m.

(c) 27.00 m.

(d) We cannot answer this question with the information given.

10. Based on the function of vertical displacement magnitude vs. time derived in Solution 1-5, we can tell that the arc of the path of the ball in two dimensions is a section of a

(a) circle.

(b) parabola.

(c) hyperbola.

(d) catenary.

CHAPTER 2

Collisions, Work, Energy, and Power

The interactions among force, velocity, and acceleration, and the ways in which energy can be expended, have infinite variety. In this chapter we'll go into a little more detail concerning how moving masses behave and interact.

Collisions

When two objects strike each other because they are in relative motion and their paths intersect at the right time, a *collision* takes place. Depending on how much of the shock of impact the objects absorb, one of the following situations must theoretically occur:

- A *perfectly elastic collision*, in which none of the shock of impact is absorbed by either of the objects

- A *perfectly inelastic collision*, in which the shock of impact is entirely absorbed by one or both of the objects
- A *partially elastic collision*, in which some, but not all, of the shock of impact is absorbed by the objects

CONSERVATION OF LINEAR MOMENTUM IN COLLISIONS

Imagine two objects or particles that coast through space on straight-line paths and then strike each other. Regardless of the nature of the matter that makes up the objects, how fast they are moving, their relative directions prior to the collision, or whether they stick together or bounce off of each other, we can be sure of one thing: The composite linear (straight-line) momentum vector will be exactly the same after the collision as before. This is *conservation of linear momentum in collisions*.

The law of conservation of linear momentum applies not only to systems having two objects or particles, but to systems having any number of objects or particles—even a countless multitude! However, the law of conservation of linear momentum holds only in a *closed system*. That means the total mass of the objects in the system must remain constant, and no forces can be introduced or imposed from the outside.

PERFECTLY ELASTIC COLLISION

Consider two balls on a collision course as shown in Fig. 2-1A. The two balls have masses m_1 and m_2, and they are moving with initial velocity vectors $\mathbf{v}_{11}$ and $\mathbf{v}_{21}$, respectively. Suppose the pre-collision momentum of ball number 1 is $\mathbf{p}_{11} = m_1\mathbf{v}_{11}$, and the pre-collision momentum of ball number 2 is $\mathbf{p}_{21} = m_2\mathbf{v}_{21}$.

In a perfectly elastic collision, the colliding objects bounce apart again in such a way that all the *kinetic energy* of their pre-collision motion is transferred into kinetic energy of their post-collision motion. (We'll get into details about kinetic energy later in this chapter.) None of the kinetic energy initially possessed by the balls turns into any other energy form because of the impact. This is the only type of collision in which the kinetic energy contained in the motions of the objects is conserved.

In the situation of Fig. 2-1A, the masses of the two balls do not change, but their velocities do. According to the law of conservation of linear momentum, the post-collision sum of the momentum vectors of the two balls has the same magnitude, and points in the same direction, as it did before the collision. Let's

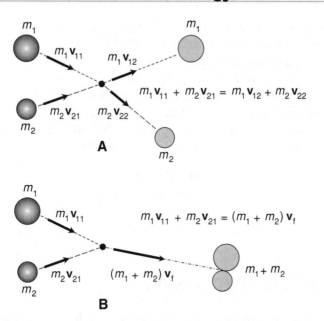

Figure 2-1   At A, an elastic or partially elastic collision. At B, a perfectly inelastic collision. For an ideal system, the total momentum before the collision is always the same as the total momentum after the collision.

call this unwavering vector **p**. Let $\mathbf{v}_{12}$ be the post-collision velocity vector for ball number 1, and let $\mathbf{v}_{22}$ be the post-collision velocity vector for ball number 2. The post-collision momentum of ball number 1 is $\mathbf{p}_{12} = m_1\mathbf{v}_{12}$, and the post-collision momentum of ball number 2 is $\mathbf{p}_{22} = m_2\mathbf{v}_{22}$. Then we can be certain of the following:

$$\begin{aligned}
\mathbf{p} &= \mathbf{p}_{11} + \mathbf{p}_{21} \\
&= \mathbf{p}_{21} + \mathbf{p}_{22} \\
&= m_1\mathbf{v}_{11} + m_2\mathbf{v}_{21} \\
&= m_1\mathbf{v}_{12} + m_2\mathbf{v}_{22}
\end{aligned}$$

PERFECTLY INELASTIC COLLISION

Now imagine two balls made of different stuff. They have the same masses and the same initial velocity vectors as in the previous case. But instead of bouncing off each other, they stick together, traveling as a single object after the collision. That means $\mathbf{v}_{12} = \mathbf{v}_{22}$. To avoid ambiguity, let's use the name $\mathbf{v}_f$ for the post-collision velocity vector of both balls, so $\mathbf{v}_f = \mathbf{v}_{12} = \mathbf{v}_{22}$. This state of affairs (Fig. 2-1B) is an example of a perfectly inelastic collision. All of the shock of impact is absorbed by one or both of the balls.

Despite the qualitative difference between this collision and the one shown in Fig. 2-1A, the law of conservation of linear momentum still holds. The pre-collision system momentum vector has the same magnitude and direction as the post-collision system momentum vector. Again, let's call it $\mathbf{p}$. As long as there are no outside forces involved, we can be sure of the following:

$$
\begin{aligned}
\mathbf{p} &= \mathbf{p}_{11} + \mathbf{p}_{21} \\
&= m_1\mathbf{v}_{12} + m_2\mathbf{v}_{22} \\
&= m_1\mathbf{v}_f + m_2\mathbf{v}_f \\
&= (m_1 + m_2)\,\mathbf{v}_f
\end{aligned}
$$

PARTIALLY ELASTIC COLLISION

Perfectly elastic collisions don't occur in the real world if the definition is taken with absolute rigor. Colliding objects always absorb some of the shock of impact, converting at least a little of their initial kinetic energy to heat, physical deformation of the objects, sound waves, or radiation. As long as the two objects don't stick together after the collision, a real-world episode can be considered partially elastic.

It's possible in theory to have a "degree of elasticity" value that can range from 0 to 1 or from 0% to 100%, where 0% represents a perfectly inelastic collision and 100% represents a perfectly elastic collision. We won't attempt to make up a quantitative definition of this term here. Your imagination should be good enough! The important fact to remember is that linear momentum is conserved in all collisions, no matter what the "degree of elasticity."

PROBLEM 2-1
Consider two balls that undergo a collision and bounce off each other. Ball number 1 has a mass of 100 g. Ball number 2 has a mass of 200 g. The pre-collision velocity of ball number 1 is 30.0 m/s, due east. The pre-collision velocity of ball number 2 is 15.0 m/s, due north. The post-collision velocity of ball number 1 is 30.0 m/s, due north. What is the total system momentum before and after the collision? What is the post-collision velocity of ball number 2? Express all values to three significant figures.

SOLUTION 2-1
Let's assign variable names to all these input values. To define the vectors, let's use a rectangular coordinate system where the positive x axis points due east and

the positive y axis points due north (Fig. 2-2A). Let's also convert grams to kilograms. Then:

$$m_1 = 100 \text{ g} = 0.100 \text{ kg}$$
$$m_2 = 200 \text{ g} = 0.200 \text{ kg}$$
$$\mathbf{v}_{11} = (x_{11}, y_{11}) = (30.0, 0.00)$$
$$\mathbf{v}_{21} = (x_{21}, y_{21}) = (0.00, 15.0)$$

This means the two balls have momentum vectors as follows:

$$\mathbf{p}_{11} = m_1 \mathbf{v}_{11}$$
$$= 0.100 \times (30.0, 0.00)$$
$$= (3.00, 0.00)$$

$$\mathbf{p}_{21} = m_2 \mathbf{v}_{21}$$
$$= 0.200 \times (0.00, 15.0)$$
$$= (0.00, 3.00)$$

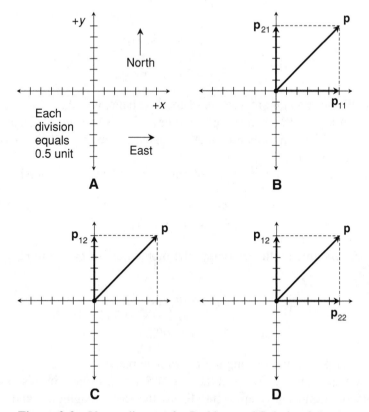

Figure 2-2   Vector diagram for Problem and Solution 2-1.

The sum of these two momentum vectors is the combined momentum **p** of the two balls:

$$\mathbf{p} = (3.00, 0.00) + (0.00, 3.00)$$
$$= (3.00, 3.00)$$

To determine the direction angle of **p**, use the formula for direction angles of vectors in rectangular coordinates that you learned in Chapter 1. Letting $x = 3.00$ and $y = 3.00$, we have:

$$\text{dir } \mathbf{p} = \arctan y/x$$
$$= \arctan (3.00/3.00)$$
$$= \arctan 1.00$$
$$= 45.0°$$

That's northeast. The magnitude $|\mathbf{p}|$ can be found using the formula for vector magnitude in rectangular coordinates that you learned in Chapter 1:

$$|\mathbf{p}| = (x^2 + y^2)^{1/2}$$
$$= [3.00^2 + 3.00^2]^{1/2}$$
$$= (9.00 + 9.00)^{1/2}$$
$$= 18.0^{1/2}$$
$$= 4.24 \text{ kg} \cdot \text{m/s}$$

The pre-collision momentum vectors of the two balls, and their sum, are shown graphically in Fig. 2-2B. According to the law of conservation of linear momentum, the post-collision momentum vector is same as the pre-collision momentum vector, or 4.24 kg · m/s.

We are told the mass and velocity of ball number 1 after the collision:

$$m_1 = 0.100 \text{ kg}$$
$$\mathbf{v}_{12} = (0.00, 30.0)$$

That is 30.0 m/s, due north. Therefore, the momentum of ball number 1 after the collision is:

$$\mathbf{p}_{12} = m_1\mathbf{v}_{12}$$
$$= 0.100 \times (0.00, 30.0)$$
$$= (0.00, 3.00)$$

Plotting **p** and $\mathbf{p}_{12}$ on the rectangular coordinate plane (Fig. 2-2C), we can envision how we can solve for $\mathbf{p}_{22}$ and thereby derive $\mathbf{v}_{22}$. In fact, in this case vector $\mathbf{p}_{22}$ should be immediately apparent! It has the same magnitude and the same direction (by coincidence) as vector $\mathbf{p}_{11}$ had before the collision, as shown in Fig. 2-2D. We solve for $\mathbf{v}_{22}$ in the form (x_{22}, y_{22}) as follows:

$$\mathbf{p}_{22} = m_2 \mathbf{v}_{22}$$
$$(3.00, 0.00) = 0.200 \times (x_{22}, y_{22})$$
$$(x_{22}, y_{22}) = (3.00, 0.00) / 0.200$$
$$= (15.0, 0.00)$$

The post-collision velocity vector of ball number 2, $\mathbf{v}_{22}$, is 15.0 m/s due east.

Figure 2-3 is a "before-and-after" drawing of this situation showing the paths of the balls in a horizontal plane. This particular scenario is a coincidence because the balls change course at perfect 90° angles. Usually, in the real world, that wouldn't be the case.

PROBLEM 2-2

Imagine two lumps of clay on a collision course. Lump number 1 masses 3.00 kg and lump number 2 masses 2.00 kg. The velocity vector of lump number 1 before the collision is 20 centimeters per second (cm/s) due south. The velocity vector of lump number 2 before the collision is 40 cm/s due east. When the two

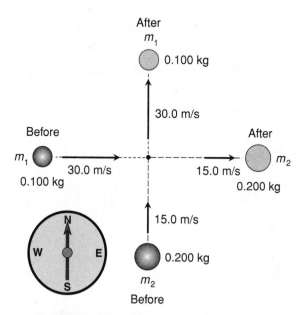

Figure 2-3   Pictorial diagram for Problem and Solution 2-1.

lumps collide, they stick together. Determine the final momentum and velocity vectors for the composite lump after the collision. Express all values to three significant figures.

SOLUTION 2-2

Once again, let's use a rectangular coordinate system where the positive x axis points due east and the positive y axis points due north. That means the negative x axis points due west and the negative y axis points due south (Fig. 2-4A). Before the collision, the velocity vector of lump number 1 points along the negative y axis and the velocity vector of lump number 2 points along the positive x axis. We convert the velocities to meters per second. Then the values, using the same variable designators as we have in the past, are:

$$m_1 = 3.00 \text{ kg}$$
$$m_2 = 2.00 \text{ kg}$$
$$\mathbf{v}_{11} = (x_{11}, y_{11}) = (0.00, -0.200)$$
$$\mathbf{v}_{21} = (x_{21}, y_{21}) = (0.400, 0.00)$$

The two lumps have pre-collision momentum vectors as follows:

$$\begin{aligned}
\mathbf{P}_{11} &= m_1 \mathbf{v}_{11} \\
&= 3.00 \times (0.00, -0.200) \\
&= (0.00, -0.600)
\end{aligned}$$

$$\begin{aligned}
\mathbf{P}_{21} &= m_2 \mathbf{v}_{21} \\
&= 2.00 \times (0.400, 0.00) \\
&= (0.800, 0.00)
\end{aligned}$$

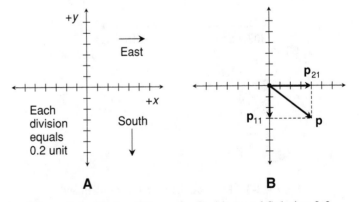

Figure 2-4   Vector diagram for Problem and Solution 2-2.

The sum of these vectors is the combined momentum vector **p** of the two balls:

$$\mathbf{p} = (0.00, -0.600) + (0.800, 0.00)$$
$$= (0.800, -0.600)$$

To determine the direction angle of **p**, recall the formula for direction angles of vectors in rectangular coordinates from the previous chapter. Letting $x = 0.800$ and $y = -0.600$, we have

$$\text{dir } \mathbf{p} = \arctan y/x$$
$$= \arctan (-0.600/0.800)$$
$$= \arctan -0.750$$
$$= -36.9°$$

This is the equivalent of 323.1°, because it is customary to keep vector direction angle measures nonnegative and less than 360°. The magnitude $|\mathbf{p}|$ can be found using the formula for vector magnitude from the previous chapter:

$$|\mathbf{p}| = (x^2 + y^2)^{1/2}$$
$$= [0.800^2 + (-0.600)^2]^{1/2}$$
$$= (0.640 + 0.360)^{1/2}$$
$$= 1.00^{1/2}$$
$$= 1.00 \text{ kg} \cdot \text{m/s}$$

The pre-collision momentum vectors of the two lumps, and their sum, are shown graphically in Fig. 2-4B.

The law of conservation of linear momentum tells us that the composite lump after the collision must have momentum vector **p**, the same as the vector sum of the pre-collision momentums. Assuming no clay breaks or flies off of the lumps during the collision, the mass m_f of the post-collision or "final" object is equal to the sum of the masses of the pre-collision lumps:

$$m_f = m_1 + m_2$$
$$= 3.00 + 2.00$$
$$= 5.00 \text{ kg}$$

The direction of the post-collision velocity vector, dir $\mathbf{v}_f$, is the same as the direction of the post-collision momentum vector, **p**. In order to obtain the post-collision velocity vector magnitude, $|\mathbf{v}_f|$, we divide the post-collision momentum vector magnitude by the post-collision mass:

$$|\mathbf{v}_f| = |\mathbf{p}| / m_f$$
$$= 1.00 / 5.00$$
$$= 0.200 \text{ m/s}$$

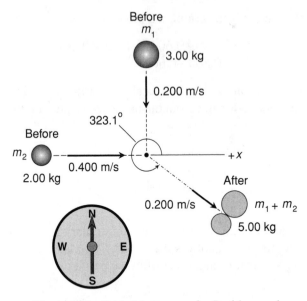

Figure 2-5   Pictorial diagram for Problem and Solution 2-2.

The post-collision velocity vector of the composite lump, $\mathbf{v}_f$, is therefore 0.200 m/s at an angle of 323.1°, or generally east-southeast. Figure 2-5 is a before-and-after drawing of this situation showing the paths of the lumps in a horizontal plane.

Work

In physics, the term *work* refers to a specific force applied over a specific distance. The standard unit of work is the *newton-meter* (N · m). In terms of *base units* in the International System, a newton-meter is equivalent to a kilogram meter squared per second squared (kg · m²/s²).

WORK IS A SCALAR

The amount of work w done on an object by the application of a constant force $\mathbf{F}$ over a defined linear displacement $\mathbf{q}$ is equal to the dot product of the force and displacement vectors:

$$w = \mathbf{F} \cdot \mathbf{q}$$

where w is in newton-meters $(N \cdot m)$, $|\mathbf{F}|$ is in newtons, and $|\mathbf{q}|$ is in meters. Work is a scalar quantity that is positive if the angle between $\mathbf{F}$ and $\mathbf{q}$ is at least 0° but less than 90°, zero if the angle between $\mathbf{F}$ and $\mathbf{q}$ is exactly 90°, and negative if the angle between $\mathbf{F}$ and $\mathbf{q}$ is greater than 90° up to a maximum possible value of 180°.

We can simplify the above to a straight scalar formula when the applied force vector maintains a constant magnitude and direction over time, and occurs in the same direction as the displacement:

$$w = |\mathbf{F}|\,|\mathbf{q}|$$

PROBLEM 2-3

Suppose you lift a 23.75-kg block for a distance of 5.000 m in a straight, vertical line using a rope and pulley system such as that shown in Fig. 2-6. How much work is expended? Express the answer to four significant figures. Assume the pulley is frictionless and the rope is massless.

SOLUTION 2-3

The force vector $\mathbf{F}$ imposed on the block occurs directly upwards and operates against the acceleration of gravity. The force and displacement occur in the same direction, so we can use the scalar formula for force in terms of mass and acceleration:

$$|\mathbf{F}| = m\,|\mathbf{a}|$$

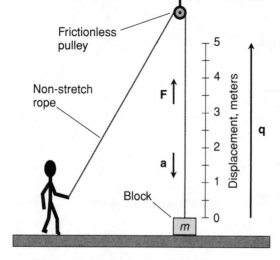

Figure 2-6   Illustration for Problem and Solution 2-3.

Let's consider the acceleration of gravity more precisely than we did in the previous chapter. The exact value of this so-called constant varies slightly from place to place on the earth's surface. A decent (although admittedly debatable) value, accurate to four significant figures, is 9.807 m/s². Let's use that value here. In this scenario, then, we have:

$$|\mathbf{F}| = 23.75 \times 9.807$$
$$= 232.9 \text{ N}$$

This is the force necessary to lift the block against gravity. The displacement vector magnitude, $|\mathbf{q}|$, is 5.000 m. Therefore:

$$w = |\mathbf{F}| \, |\mathbf{q}|$$
$$= 232.9 \times 5.000$$
$$= 1165 \text{ N} \cdot \text{m}$$

PROBLEM 2-4

Suppose you lift the same 23.75-kg block using the same pulley system as in the previous scenario, but this time you haul it up an inclined plane on a cart over a *slant displacement* of 5.000 m, as shown in Fig. 2-7. The plane is slanted at 30° relative to the vertical. The cart has frictionless wheels. How much work is done? Express the answer to four significant figures.

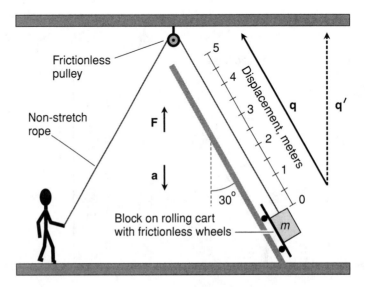

Figure 2-7 Illustration for Problem and Solution 2-4.

SOLUTION 2-4

The force vector **F** responsible for doing work still occurs directly upwards, against the acceleration of gravity. (No force component is necessary to move the block in the horizontal direction.) That means $|\mathbf{F}|$ is the same as before:

$$\begin{aligned}|\mathbf{F}| &= m\,|\mathbf{a}|\\ &= 23.75 \times 9.807\\ &= 232.9 \text{ N}\end{aligned}$$

The displacement that manifests useful work is the distance that the block travels in an upward direction. That is the vertical component of the slant displacement vector **q**. Call this vertical component vector **q′**. The magnitude of **q′** is equal to the magnitude of **q** times the cosine of the angle θ between **q** and **q′**, as follows:

$$|\mathbf{q'}| = |\mathbf{q}| \cos \theta$$

The work is calculated according to this formula:

$$\begin{aligned}w &= |\mathbf{F}|\,|\mathbf{q'}|\\ &= |\mathbf{F}|\,|\mathbf{q}| \cos \theta\\ &= \mathbf{F} \cdot \mathbf{q}\end{aligned}$$

In the example of Fig. 2-7, $\theta = 30°$. Therefore:

$$\begin{aligned}w &= |\mathbf{F}|\,|\mathbf{q}| \cos 30°\\ &= 232.9 \times 5.000 \times 0.8660\\ &= 1008 \text{ N} \cdot \text{m}\end{aligned}$$

WORK DONE BY A VARIABLE FORCE

Sometimes it is necessary to calculate work when the force varies with displacement. Examples are electric and magnetic fields between charged or magnetized objects. We may also encounter this type of situation when an object travels in a variable gravitational field, such as from earth to the moon.

Let s represent the separation between two objects, one movable and the other fixed. Let s_1 be the initial separation, and let s_2 be the separation after a certain amount of work has been done to get the movable object from one point to another. Let **F** be the force vector applied to the movable object. Let $|\mathbf{F}|(s)$ represent the applied force vector magnitude as a function of the separation between the objects. Suppose that the applied force vector **F** and the displacement vector **q** always point in exactly the same direction, and the movable object travels in a straight line. The work w required to change the separation from s_1 to s_2 is given by this definite integral:

$$w = \int_{s_1}^{s_2} |\mathbf{F}|(s)\, ds$$

PROBLEM 2-5

Imagine a pair of electrostatically charged objects on a horizontal surface. The objects have opposite, constant, equal electrical charges, so they attract each other. One object is anchored down, and the other can be moved on a mini-cart with frictionless wheels, as shown in Fig. 2-8. The force of attraction between two charged bodies is inversely proportional to the square of the separation between the charge centers. Let $\mathbf{F}$ be the force we must apply to the object on the mini-cart to make it move toward the left. Let $|\mathbf{F}|$ be the magnitude of that applied force in newtons. Let s be the separation between the charge centers in centimeters. Suppose the total charge quantity is such that the applied force vector magnitude, as a function of the separation between charge centers, is given by this formula:

$$|\mathbf{F}|(s) = 2.55\,/\,s^2$$

Now imagine that you use a little stick to roll the mini-cart against the force of electrostatic attraction from $s_1 = 3.00$ cm to $s_2 = 10.00$ cm. How much work is done? Note that because the displacement is horizontal, gravity has no effect. Express the answer to three significant figures.

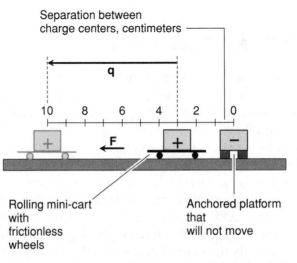

Figure 2-8   Illustration for Problem and Solution 2-5.

SOLUTION 2-5

Plugging the numbers into the formula for the work done by a variable force, we get this:

$$w = \int_{s_1}^{s_2} |\mathbf{F}|(s) \, ds$$

$$= \int_{3.00}^{10.00} \frac{2.55}{s^2} \, ds$$

We should not convert centimeters to meters here, because we're explicitly given the formula for force in newtons based on the separation in centimeters. To evaluate the above expression, we must know the antiderivative of $|\mathbf{F}|(s)$. Let's call it $|\mathbf{F}|*(s)$. If you don't remember how to integrate a function of this form, you can use a table of indefinite integrals to find the antiderivative:

$$|\mathbf{F}|*(s) = -2.55 \,/\, s$$

We evaluate $|\mathbf{F}|*(s)$ from $s_1 = 3.00$ to $s_2 = 10.00$ to determine the work done:

$$\begin{aligned} w &= (-2.55 \,/\, s_2) - (-2.55 \,/\, s_1) \\ &= (-2.55 \,/\, 10.00) + (2.55 \,/\, 3.00) \\ &= -0.255 + 0.850 \\ &= 0.595 \; \mathrm{N \cdot m} \end{aligned}$$

Energy

Energy is closely associated with work. When work is done on an object, that object acquires "latent" energy that can be used to do mechanical work at a later time; it can also be converted into some other form of energy such as heat or electricity.

UNITS OF ENERGY

Recall that if we break down the newton-meter ($\mathrm{N \cdot m}$) into base units, we get a kilogram meter squared per second squared ($\mathrm{kg \cdot m^2/s^2}$). That's also known as a *joule*, symbolized by the uppercase letter J. The joule is the fundamental unit of energy. Small amounts of energy can be expressed in *millijoules* (mJ), *microjoules*, (μJ), *nanojoules* (nJ), or *picojoules* (pJ). Large amounts of energy can be expressed in terms of *kilojoules* (kJ), *megajoules* (MJ), *gigajoules* (GJ), or *terajoules* (TJ). By now you should know what these prefix multipliers mean!

POTENTIAL ENERGY

When an object with mass m is moved by a force $\mathbf{F}$ over a displacement $\mathbf{q}$, work is done on that object. For example, if you lift a stone straight up, you do work to get it from the lower altitude to the higher. You must do work to compress a spring, stretch a rubber band, or pull two oppositely charged objects apart. Imagine what happens if you let a stone drop, release a compressed spring, let go of a stretched-out rubber band, or let two oppositely charged objects have their way. The stone falls; the spring expands; the rubber band contracts; the charged objects fly at each other. The work you have done is "undone" with visible—perhaps violent—results.

Suppose you lift a 5-kg stone above a dry, sandy beach. If you raise the stone a couple of centimeters and then let it drop, it will strike the sand without much effect. If you raise it a couple of meters, it will throw sand out and create a little depression. If you loft the stone 10 km and then let it drop, it will make a significant crater; hopefully the impact won't occur where it can damage anything or hurt anyone! The falling of a heavy object can be put to constructive use, such as pounding a pole into the ground. It can also do damage, as old-world warriors with giant slingshots discovered when they hurled boulders into fortified cities. There is something about moving an object from one place to another against a force field that imparts to that object a pent-up form of energy known as *potential energy*.

Suppose a force $\mathbf{F}$ is applied to an object against gravity, the tension of a spring or rubber band, the electrostatic attraction between two charged objects, or some other force field. Suppose the object is moved over a displacement $\mathbf{q}$. Then the potential energy, E_{p}, that the object acquires is given by:

$$E_{\mathrm{p}} = \mathbf{F} \cdot \mathbf{q}$$

If $|\mathbf{F}|$ is in newtons and $|\mathbf{q}|$ is in meters, then E_{p} is in newton-meters, just like work. But when we consider it as energy, it is customary to express it in joules.

Potential energy is a scalar quantity. As is the case with work, we can use a scalar formula when the applied force vector maintains a constant magnitude and direction over time, and occurs in the same direction as the displacement vector:

$$E_{\mathrm{p}} = |\mathbf{F}|\,|\mathbf{q}|$$

If $\mathbf{F}$ is constant over time but occurs in a different direction from $\mathbf{q}$, and if θ is the angular difference between dir $\mathbf{F}$ and dir $\mathbf{q}$, then:

$$E_{\mathrm{p}} = |\mathbf{F}|\,|\mathbf{q}|\cos\theta$$

If $\mathbf{F}$ occurs in the same direction as $\mathbf{q}$ but varies in magnitude, we have a situation similar to that with work done by a variable force. Let s represent the separation between two objects, one movable and the other fixed. Let s_1 be the initial separation between the objects, and let s_2 be the final separation. Let $|\mathbf{F}|(s)$ represent $|\mathbf{F}|$ as a function of s. The potential energy E_p imparted to the movable object is given by:

$$E_p = \int_{s_1}^{s_2} |\mathbf{F}|(s)\, ds$$

KINETIC ENERGY IN TERMS OF FORCE AND DISPLACEMENT

Let's revisit the scenario of Problem 2-3 and Fig. 2-6. Suppose you let go of the rope after having lifted the 23.75-kg block 5.000 m. You let the block fall back to the surface. The block starts moving as gravity imposes a force vector $\mathbf{G}$ on it. You gave a certain amount of potential energy E_p to the block when you lifted it. That potential energy (in joules) was equal, in terms of kilogram meters squared per second squared, to the work you did (in newton-meters). When the block falls, its potential energy is converted to *kinetic energy*. Just as the block makes impact, all the work you did against gravity has been "undone" by gravity. The kinetic energy E_k of the block at impact is equal to the potential energy E_p that was imparted to the block when you hoisted it.

The vector formula for the kinetic energy imparted to an object acted upon by a constant force $\mathbf{G}$ over a displacement $\mathbf{r}$ is:

$$E_k = \mathbf{G} \cdot \mathbf{r}$$

where $\mathbf{G}$ is the force and $\mathbf{r}$ is the displacement. For a freely falling object near the earth's surface, $|\mathbf{G}|$ is constant and both $\mathbf{G}$ and $\mathbf{r}$ point straight down, so we can use the following scalar formula:

$$E_k = |\mathbf{G}|\,|\mathbf{r}|$$

Suppose a force $\mathbf{G}$ having constant direction but variable magnitude is applied to an object, causing that object to start moving from a dead stop and travel in a straight line. Let $|\mathbf{G}|(s)$ represent $|\mathbf{G}|$ as a function of the object's position s. The kinetic energy E_k that the object possesses after traveling from an initial point s_1 to a final point s_2 is given by:

$$E_k = \int_{s_1}^{s_2} |\mathbf{G}|(s)\, ds$$

Kinetic energy, like potential energy, is expressed in joules.

KINETIC ENERGY IN TERMS OF MASS AND SPEED

There is way to express the instantaneous value of E_k for an object that moves in a straight line, based on its mass m and its instantaneous velocity vector magnitude $|\mathbf{v}|$. That formula is:

$$E_k = m\,|\mathbf{v}|^2 / 2$$

This formula can come in handy when the magnitude of the applied force, or the distance an object travels, is not known, but the object's mass and speed are. The expression $(m\,|\mathbf{v}|^2 / 2)$ can be substituted for E_k on the left-hand side of any of the force-displacement equations from the previous paragraph.

PROBLEM 2-6

Once again, consider the block-and-pulley scenario. How much kinetic energy does the 23.75-kg block have after we let the rope go (Fig. 2-9) so the block falls freely for 5.000 m back to the surface? Express the answer to four significant figures, for the instant just before impact.

SOLUTION 2-6

Remember that when a mass is acted on by a force, causing the mass to accelerate in the direction of that force, then the force vector magnitude is equal to

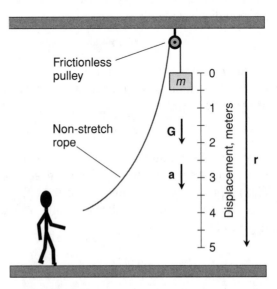

Figure 2-9   Illustration for Problems and Solutions 2-6 and 2-7.

the product of the mass and the acceleration vector magnitude. Here, the mass is 23.75 kg and the magnitude of the acceleration of gravity is 9.807 m/s^2, so the force vector magnitude $|\mathbf{G}|$ is 23.75 × 9.807 = 232.9 N. The kinetic energy E_k that the block possesses after falling through a displacement $\mathbf{r}$ straight down is equal to the product of the force vector magnitude and the displacement vector magnitude:

$$E_k = |\mathbf{G}|\,|\mathbf{r}|$$

In this example, $|\mathbf{r}| = 5.000$ m. Therefore:

$$E_k = 232.9 \times 5.000$$
$$= 1165 \text{ J}$$

PROBLEM 2-7
How fast is the block in the scenario of Problem 2-6 and Fig. 2-9 moving when it hits the surface? Derive the answer in two different ways to four significant figures.

SOLUTION 2-7
We know that E_k = 1165 J. We also know that m = 23.75 kg. We can plug these numbers into the formula for kinetic energy in terms of mass and instantaneous velocity vector magnitude. In that formula, the denominator is exactly 2, so we can express it to as many significant figures as we need by adding zeros after the decimal point. Therefore:

$$E_k = m\,|\mathbf{v}|^2 / 2.000$$
$$1165 = 23.75 \times |\mathbf{v}|^2 / 2.000$$
$$|\mathbf{v}|^2 = 2.000 \times 1165 / 23.75$$
$$|\mathbf{v}|^2 = 98.11$$
$$|\mathbf{v}| = 9.905 \text{ m/s}$$

The other way to find $|\mathbf{v}|$ is rather messy, but it works. Let $|\mathbf{a}|(t)$ be the function that defines the magnitude $|\mathbf{a}|$ of the acceleration of gravity as a function of elapsed time t. This is a constant function in the vicinity of the earth's surface, assuming there is no air resistance:

$$|\mathbf{a}|(t) = 9.807$$

The function $|\mathbf{v}|(t)$, defining the downward velocity magnitude $|\mathbf{v}|$ of a falling object in terms of the elapsed time t, is found by integrating $|\mathbf{a}|(t)$ with respect to time:

$$|\mathbf{v}|(t) = \int |\mathbf{a}|(t)\,dt$$
$$= \int 9.807\,dt$$
$$= 9.807t + c_1$$

The function $|\mathbf{r}|(t)$, denoting the downward displacement magnitude $|\mathbf{r}|$ of a falling object in terms of the elapsed time t, is found by integrating $|\mathbf{v}|(t)$ with respect to time:

$$|\mathbf{r}|(t) = \int |\mathbf{v}|(t)\, dt$$
$$= \int 9.807t\, dt$$
$$= 4.904t^2 + c_2$$

Let's set the starting time at $t = 0$, eliminating the constants of integration. We can plug in $|\mathbf{r}|(t) = 5.000$ to the above formula find out how long it takes the block to fall:

$$5.000 = 4.904t^2$$
$$t^2 = 5.000 / 4.904$$
$$t = 1.010 \text{ s}$$

Now we can go back to the function $|\mathbf{v}|(t)$ and plug in the value $t = 1.010$, obtaining the velocity vector magnitude at impact:

$$|\mathbf{v}| = 9.807 \times 1.010$$
$$= 9.905 \text{ m/s}$$

PROBLEM 2-8

After the block in the above scenario has hit the surface and come to rest, its kinetic energy must be zero according to the mass-velocity formula, because its velocity is zero. What happens to all that kinetic energy?

SOLUTION 2-8

It is converted into other forms of energy such as noise, heat, earth vibration, and maybe even some flying debris! The block might bounce a couple of times, so the conversion of kinetic energy to these other forms might take a second or two.

THE WORK-ENERGY THEOREM

The total amount of work, w_t, done on an object can be manifested as a change in its kinetic energy. This rule of mechanics is known as the *work-energy theorem*, stated mathematically as follows:

$$w_t = \Delta E_k$$

where the symbol Δ stands for "the change in."

MECHANICAL ENERGY

The sum of the potential energy and the kinetic energy that an object or system contains is known as the *mechanical energy*, symbolized E_m. Thus:

$$E_m = E_p + E_k$$

We can expand on the work-energy theorem in complex systems having both potential energy and kinetic energy:

$$w_t = \Delta E_m = \Delta E_p + \Delta E_k$$

This holds true provided no external force or event affects the object's motion or position. Consider the following examples.

- You lift a block straight up by doing 250 N · m of work on it against the force of gravity. Therefore, the block gains 250 J of potential energy. You set the block down again, doing −250 N · m of work on it. Therefore, it loses 250 J of potential energy.

- You release a stone from a slingshot after you have done 76.88 N · m of work to stretch the elastic band. Therefore, the stone leaves the slingshot with an initial kinetic energy of 76.88 J. In a theoretically ideal situation such as a gravitation-free vacuum, the stone maintains E_k = 76.88 J until or unless it hits something or is acted on by an outside force.

- In the real world, after you let the stone fly out from the slingshot, it gains kinetic energy over and above the initial 76.88 J as it accelerates in a downward arc because of the external force of gravity. Ultimately, the external friction caused by air resistance limits the maximum kinetic energy the stone can attain, or else the stone hits the surface and its kinetic energy is converted to some other form of energy.

PROBLEM 2-9

Suppose you pull the movable object in the scenario of Problem 2-5 away from the fixed object until the charge centers are 10.00 cm apart. You let the movable object go, and it accelerates toward the fixed object because of electrostatic attraction. Suppose the total charge quantity is such that the attractive force vector magnitude $|\mathbf{G}|$, as a function of the separation s between charge centers, is given by this formula:

$$|\mathbf{G}|(s) = 2.55 \, / \, s^2$$

This is the same electrostatic charge situation as described in Problem 2-5 and Fig. 2-8. How much kinetic energy, E_k, does the movable object possess after it

Advanced Physics Demystified

has traveled 5.00 cm as shown in Fig. 2-10? Assume it starts from a dead stop. Express the answer to three significant figures.

SOLUTION 2-9

Note that $|\mathbf{G}|$ in this situation is equal to $|\mathbf{F}|$ in the scenario of Problem 2-5, although the vectors $\mathbf{G}$ and $\mathbf{F}$ point in opposite directions. Note also that $\mathbf{q}$ and $\mathbf{r}$ lie along the same straight line, although they point in opposite directions. According to the work-energy theorem, the kinetic energy the movable object possesses in this situation is equal to the amount of work w it would take to pull it away from the fixed object from $|\mathbf{q}| = 5.00$ cm to $|\mathbf{q}| = 10.00$ cm. We can find w for this case just as we did in Solution 2-5, but using different limits for the definite integral:

$$w = \int_{5.00}^{10.00} \frac{2.55}{s^2}\, ds$$

Recall the antiderivative function $|\mathbf{F}|*(s)$ from Solution 2-5:

$$|\mathbf{F}|*(s) = -2.55\,/\,s$$

We evaluate $|\mathbf{F}|*(s)$ from $s_1 = 5.00$ to $s_2 = 10.00$ to determine the work done:

$$
\begin{aligned}
w &= (-2.55\,/\,s_2) - (-2.55\,/\,s_1) \\
&= (-2.55\,/\,10.00) + (2.55\,/\,5.00) \\
&= -0.255 + 0.510 \\
&= 0.255\ \mathrm{N} \cdot \mathrm{m}
\end{aligned}
$$

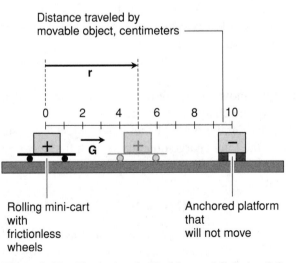

Distance traveled by movable object, centimeters

r

0 2 4 6 8 10

G

Rolling mini-cart with frictionless wheels

Anchored platform that will not move

Figure 2-10   Illustration for Problem and Solution 2-9.

The amount of kinetic energy that the movable object possess after it has traveled 5.00 cm, acted upon from a dead stop by the force **G** over the displacement **r** as shown in Fig. 2-10, is therefore $E_k = 0.255$ J.

Power

In the context of mechanics, *power* is the rate at which work is done. More generally, power is the rate at which energy is expended, dissipated, or converted to another form. Power, like energy, is a scalar quantity. As a variable in equations, it is symbolized by the uppercase, italic letter P. Power can be averaged over a period of time (P_{avg}) or considered as an instantaneous value at a point in time (P_{inst}).

UNITS OF POWER

The standard unit of power is the *joule per second* (J/s), more commonly known as the *watt* (W). In base units, a watt is the equivalent of a kilogram meter squared per second cubed (kg · m²/s³). Small amounts of power can be expressed in *milliwatts* (mW), *microwatts*, (µW), *nanowatts* (nW), or *picowatts* (pW). Large amounts of power can be expressed in terms of *kilowatts* (kW), *megawatts* (MW), *gigawatts* (GW), or *terawatts* (TW).

AVERAGE POWER

In the situation of Problem 2-3 and Fig. 2-6, a 23.75-kg block acquires $E_p = 1165$ J of potential energy as it is lifted straight up by 5.000 m. Suppose it takes you 10.00 s of time to lift that block. The *average power* you expend over that time, P_{avg}, is equal to the potential energy the block gains divided by the time taken to gain it. This is true even if you don't lift the block at a constant speed, as long as it rises 5.000 m in 10.000 s. Mathematically:

$$P_{avg} = E_p / t$$
$$= 1165 / 10.00$$
$$= 116.5 \text{ W}$$

Now imagine that the block is allowed to fall freely as described in Problems 2-6 and 2-7 and illustrated by Fig. 2-9. We have found, in the course of calculating its speed at impact, that the block takes 1.010 s to fall 5.000 m. In this time it acquires

1165 J of kinetic energy. Just prior to impact, all of the potential energy the block gained when it was lifted has been converted into motion. The average power that gravity expends on the block as it drops 5.000 m is therefore:

$$P_{avg} = E_k / t$$
$$= 1165 / 1.010$$
$$= 1153 \text{ W}$$
$$= 1.153 \text{ kW}$$

This is much greater than the average power you expended in lifting the block, because the block falls much faster.

INSTANTANEOUS POWER

The rate at which work is done is not always constant. Imagine that you carry a heavy box up a flight of stairs to move it from one storage location to another. You must do work in order to move the box to a higher level. But suppose you're not putting much effort into the task. Your boss yells, "Faster!" Rather than telling the boss off, you climb the stairs faster. This increases the rate at which you do the work, so the *instantaneous power* you expend at each point in time, P_{inst}, increases.

Let $w(t)$ represent the total work done as a function of the elapsed time t from the start of the job. Let $P_{inst}(t)$ represent the function of instantaneous power vs. time. Then:

$$P_{inst}(t) = dw(t) / dt$$

The above generalized formula holds whether the instantaneous power is constant or not. The instantaneous power can be graphically represented by the slope of the work-vs.-time curve at a specific point in time. Two examples are shown in Fig. 2-11. At A, the instantaneous power is constant, so the graph of the function $w(t)$ is a straight line, and P_{inst} is equal to the slope of the line at all points in time. At B, the instantaneous power varies, so $w(t)$ is not a straight line. In this situation, the value of $P_{inst}(t)$ at a specific point t_0 is equal to the slope of the line tangent to the curve at that point, which is the derivative of the function at that point. However, the slopes of the tangent lines at other points in time can be smaller (representing less power) or greater (representing more power) than is the case at t_0.

WORK AS ACCUMULATED POWER

There's a "flip side" to the fact that power is the rate at which work is done. Work can be expressed as the effect of accumulated power over time.

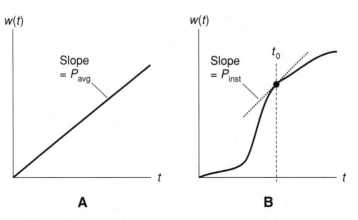

Figure 2-11 At A, the instantaneous power is constant over time. At B, the instantaneous power varies with time.

Let's go back to the box-up-the-stairs scene. You're climbing the stairs, sometimes fast (when the boss is watching) and sometimes not so fast (when the boss is not watching). The box makes progress on its way from its old storage space to its new one, sometimes rapidly, sometimes slowly. The instantaneous power, P_{inst}, that you put into lifting the box against gravity varies with the time t.

The function $w(t)$, representing accumulated work done as a function of time, can be found by integration:

$$w(t) = \int P_{inst}(t)\, dt$$

Figure 2-12 shows an example of this in graphical form. If we define the process as starting at time $t = 0$ and the initial accumulated work as $w(0) = 0$, we need not worry about constants of integration, so the accumulated work up to time t_0 can be expressed as the area under the curve (shaded region) between $t = 0$ and $t = t_0$.

PROBLEM 2-10

Suppose you stand in the gondola of a hot-air balloon hovering high over a lake. You drop a 2.200-kg gold ingot and let it fall freely. You know that the force of gravity does work on the ingot. You set $t = 0.00$ at the moment you let the ingot go, and you make sure you don't impart any velocity to it when you drop it. (That ensures that you won't have to deal with constants of integration when doing the calculations to come.) Neglecting air resistance, and taking the acceleration of gravity to be 10.00 m/s^2, derive functions for:

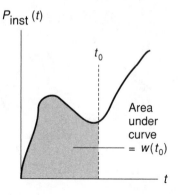

Figure 2-12 The function of accumulated work vs. time is equal to the integral of the function of instantaneous power vs. time, as shown by the shaded region.

1. The ingot's downward acceleration vector magnitude vs. time, $|\mathbf{a}|(t)$
2. The ingot's downward velocity vector magnitude vs. time, $|\mathbf{v}|(t)$
3. The ingot's downward displacement vector magnitude vs. time, $|\mathbf{q}|(t)$
4. The kinetic energy possessed by the ingot vs. time, $E_k(t)$
5. The accumulated work done by gravity on the ingot vs. time, $w(t)$
6. The instantaneous power inherent in the ingot vs. time, $P_{\text{inst}}(t)$
7. The average power inherent in the ingot during the first 1.000 s of its descent

SOLUTION 2-10

1. The downward acceleration vector magnitude is constant, and is equal to the acceleration of gravity. Therefore, for values in meters per second squared:

$$|\mathbf{a}|(t) = 10.00$$

2. The function $|\mathbf{v}|(t)$ is the indefinite integral of $|\mathbf{a}|(t)$ with respect to t. The constant of integration is 0. Therefore, for values in meters per second:

$$|\mathbf{v}|(t) = 10.00t$$

3. The function $|\mathbf{q}|(t)$ is the indefinite integral of $|\mathbf{v}|(t)$ with respect to t. The constant of integration is exactly equal to 0. Therefore, for values in meters:

$$|\mathbf{q}|(t) = 5.000t^2$$

4. The function $E_k(t)$ can be derived from the formula for kinetic energy in terms of velocity vector magnitude:

$$E_k = m \, |\mathbf{v}|^2 / 2$$

We have found that $|\mathbf{v}|(t) = 10.00t$, and we have been told that $m = 2.200$, so we can substitute $10.00t$ for $|\mathbf{v}|$ in this formula and plug in the known value for m, obtaining the following function for values in joules:

$$
\begin{aligned}
E_k(t) &= 2.200 \times (10.00t)^2 / 2.000 \\
&= 2.200 \times 100.0t^2 / 2.000 \\
&= 110.0t^2
\end{aligned}
$$

5. From the work-energy theorem, and from the fact that $E_k = 0.000$ at $t = 0.000$, we can conclude that the function $w(t)$ is the same as $E_k(t)$. Therefore, for values in newton-meters:

$$w(t) = 110.0t^2$$

6. The function $P_{inst}(t)$ can be derived from $w(t)$ by differentiation, giving us the value in watts:

$$
\begin{aligned}
P_{inst}(t) &= dw(t) / dt \\
&= 220.0t
\end{aligned}
$$

7. The average power P_{avg} inherent in the ingot during the first second of its descent can be found as follows for $t = 1.000$:

$$
\begin{aligned}
P_{avg} &= E_k(t) / t \\
&= 110.0t^2 / t \\
&= (110.0 \times 1.000^2) / 1.000 \\
&= 110.0 \text{ W}
\end{aligned}
$$

Quiz

This is an "open book" quiz. You may refer to the text in this chapter. A good score is 8 correct. Answers are in the back of the book.

1. Imagine that you're driving a 1000-kg car with frictionless wheels along a frictionless, perfectly straight, perfectly horizontal and level highway. There is no wind or air resistance. Suddenly you run out of gas. You know you can coast indefinitely, but you also know you'll get into an accident if you do that. So you apply the brakes and come to a stop, grab your cell phone, and hope that it will work so you can call for help. In the process of stopping the car, it travels over a horizontal distance of 75.00 m. How

much work is expended *against gravity* to get from full speed to a full stop in this situation? Consider the acceleration of gravity to be 9.807 m/s². Express the answer to four significant figures.

(a) 0.000 N · m.

(b) 7.355 × 10⁵ N · m.

(c) −7.355 × 10⁵ N · m.

(d) It depends on the initial speed of the car.

2. Suppose a 10.00-g bullet is fired out of a gun at a speed of 1.200 km/s. What is its kinetic energy at the instant it leaves the gun muzzle?

(a) 12.00 J.

(b) 144.0 J.

(c) 7.200 kJ.

(d) 14.40 kJ.

3. Suppose two lumps of clay approach each other from opposite directions in deep space. Lump number 1 masses 4.00 kg and travels at 5.00 m/s. Lump number 2 masses 10.0 kg. When the lumps collide, no material flies off. The lumps stick together, and the resulting object is stationary relative to you, the observer. How fast was lump number 2 moving before the collision relative to you, the observer?

(a) 40 cm/s.

(b) 50 cm/s.

(c) 2.00 m/s.

(d) 2.50 m/s.

4. When work is done on an object, it is the equivalent of

(a) a change in its displacement.

(b) a change in its mass.

(c) a change in its mechanical energy.

(d) a change in its velocity.

5. Which of these actions represents negative mechanical work?

(a) Compressing the spring in a "mechanical pitcher" to launch a baseball.

(b) Lowering a bucket on a rope down into an old-fashioned well.

(c) Pushing two magnets together against the force of repulsion when their like poles face each other.

(d) Drawing back the elastic band in a slingshot to hurl a marble at a tree.

6. According to the results in Solution 2-10, we can conclude that the instantaneous power inherent in a heavy, dense object as it falls in a constant gravitational field, neglecting air resistance or interference from any other outside force, is

   (a) constant.

   (b) directly proportional to the time elapsed after its release.

   (c) directly proportional to the square of the time elapsed after its release.

   (d) directly proportional to the square root of the time elapsed after its release.

7. You lift a 2.1-kg brick a distance of 126 cm while building a wall. Then you accidentally drop it. How much kinetic energy does the brick have in the instant before it hits the ground? Consider the acceleration of gravity to be 9.807 m/s^2.

   (a) 1.6 kJ.

   (b) 2.6 kJ.

   (c) 16 J.

   (d) 26 J.

8. Suppose mechanical work is done on an object according to the following work-vs.-time function:

$$w(t) = t^2 + 2t + 1$$

   where the work $w(t)$ is in newton-meters and the time t is in seconds. What is the instantaneous power P_{inst}, in watts, at $t = 3$ s?

   (a) 21 W.

   (b) 17 W.

   (c) 8 W.

   (d) We cannot answer this unless we know the constant of integration.

9. Imagine that you are on a space walk outside a vessel coasting through interstellar space. You have launched a reference probe that is perfectly stationary relative to the ship, equipped with precision laser velocity-measuring devices. A meteoroid approaches at 4.0 m/s, strikes the ship, and bounces off in exactly the opposite direction at 2.0 m/s. You grab the meteoroid, take it into the ship, put it into an inertial mass meter, and find out that it masses 1.0 kg. The ship's mass is 20,000 kg without you on board. You know that the impact has caused the space ship to start moving relative to the probe in the same direction the meteoroid was originally traveling. What will the

probe indicate as the magnitude of the ship's post-collision velocity vector? Express the answer in millimeters per second (mm/s).

(a) 0.30 mm/s.

(b) 0.20 mm/s.

(c) 0.10 mm/s.

(d) 0.03 mm/s.

10. Suppose that in the scenario of the previous question, you had been on board the ship instead of walking in space outside it. Your mass is 60 kg. How much of a difference in the magnitude of the ship's post-collision velocity vector, as measured by the reference probe, would your presence on board make? Express the answer in micrometers per second (μm/s).

(a) It would have been 0.30 μm/s less.

(b) It would have been 0.60 μm/s less.

(c) It would have been 0.90 μm/s less.

(d) It would not have made any difference whatsoever.

CHAPTER 3

Circular and Harmonic Motion

When an object spins or orbits at a constant rate, it has certain properties. In this chapter we'll examine the nature of *uniform circular motion*. Then we'll see how this form of motion relates to regular, back-and-forth oscillation.

Aspects of Uniform Circular Motion

Circular motion can occur as *rotation* or *revolution*. In rotation, a solid object turns around a point or line. A disk rotates around its center; the earth rotates on its axis. In revolution, an object follows a circular path around a central point. A carriage revolves around the center of a Ferris wheel; a satellite revolves around the earth. Sometimes these two terms both apply to a single phenomenon. For example, a point on the edge of a rotating disk revolves around the center of the disk.

ALTERNATIVE NOTATION FOR VECTOR MAGNITUDE

There is a fundamental difference between a vector and its magnitude. This has been rigorously emphasized in the first two chapters. For example, the magnitude of a velocity vector **v** has been denoted as $|\mathbf{v}|$, and the magnitude of a displacement vector **q** has been written $|\mathbf{q}|$. But velocity magnitude can also be written v (in italics, not boldface, and without vertical lines), and displacement magnitude can be written q. This alternative notation is often used when the vector nature of a quantity or phenomenon isn't being considered.

When we talk about *speed*, we don't think about it in terms of magnitude and direction, but as a scalar expression of how fast something is moving. When we talk about *distance*, we don't think about the direction of a displacement vector, but only about how far away one point is from another. Technically, we can state things like this:

$$\text{For velocity: } |\mathbf{v}| = v$$
$$\text{For displacement: } |\mathbf{q}| = q$$
$$\text{For acceleration: } |\mathbf{a}| = a$$
$$\text{For force: } |\mathbf{F}| = F$$

In this chapter, we'll start using this alternative notation. You should get accustomed to seeing it both ways. When a quantity or phenomenon is denoted in boldface, think of its as a vector. When it's denoted in italics, think of it as a scalar without concern about the direction or orientation. Of course, some quantities and phenomena, such as mass and energy, are scalars by nature.

RADIUS AND RADIAL DISPLACEMENT

Imagine that you stand on a flat prairie and whirl a golf ball that is attached to the end of a non-elastic, massless string. Ignore the small motions you make with your hand to keep the ball moving, and suppose the wind is not a factor either. The ball's path is a circle in a horizontal plane, as shown in Fig. 3-1A. The circle has radius r equal to the length of the string. The circumference of the circle is equal to $2\pi r$, where π (the lowercase Greek letter pi) represents a constant roughly equal to 3.14159.

Sometimes we want to specify the direction in which the radius is defined. Then we must consider the *radial displacement* vector **r** from the center point or axis to a revolving object. As uniform circular motion takes place, dir **r** constantly changes for that object, but $|\mathbf{r}|$ or r remains constant.

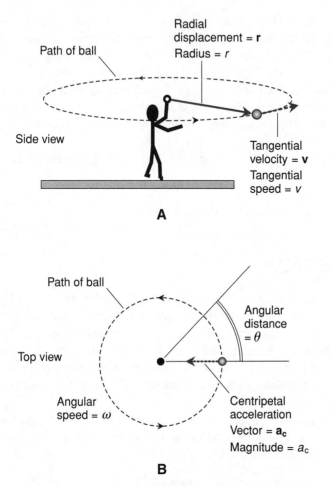

Figure 3-1   Aspects of uniform circular motion. At A, side view showing radial displacement **r**, radius r, tangential velocity **v**, and tangential speed v. At B, top view showing angular distance θ, angular speed ω, centripetal acceleration **a**$_c$, and centripetal acceleration magnitude a_c. Vector lengths are arbitrary.

ANGULAR DISTANCE

In the scenario of Fig. 3-1, the distance q that the ball travels over a span of time t can be measured around the circumference of the circle. This is not linear displacement because the path is curved! The rules of linear displacement do not apply to circular motion, except over distances that represent such a small fraction of the circle that the part of the curve under consideration is essentially straight.

In circular motion, distance can be defined according to the angle of rotation or revolution that occurs. This is called *angular distance* and is symbolized by the lowercase italic Greek letter θ (theta). Angular distance can be expressed in radians (rad), where there are 2π rad in a complete circle. It can also be expressed in angular degrees (°), where there are 360° in a perfect circle, or in one complete revolution. In physics, the radian is preferred.

Actual distance q, radius r, and angular distance θ are related as follows:

$$q = r\theta$$

where q and r are in meters, and θ is in radians.

ANGULAR SPEED

In circular motion, *angular speed* is the angle through which rotation or revolution occurs per unit time. In physics, this quantity is symbolized by the lowercase italic Greek letter ω (omega) and is expressed in radians per second (rad/s). If you're an auto mechanic or a computer hard-drive designer, you might talk about revolutions or rotations per minute (rpm) or per second (rps). Once in a while you'll hear about odd angular speed units such as radians per millisecond (rad/ms) or degrees per hour (°/h).

Angular speed ω is inversely related to the *period*, which is the time required for precisely one complete rotation or revolution. The symbol for period is an uppercase italicized letter T. Here is the formula:

$$\omega = 2\pi / T$$

where ω is in radians per second and T is in seconds.

In the situation of Fig. 3-1, suppose you whirl the ball around your head at a rate of 1 rps. Then its angular speed in radians per second is 2π times that, or about 6.28318 rad/s. This is true regardless of the length of the string. If the string is short, the ball doesn't move very fast. If you have a spool of string and you let more and more of the string out but keep the angular speed of the ball constant, it travels faster and faster through space.

The following formula expresses instantaneous angular speed in terms of angular distance and time:

$$\omega = d\theta / dt$$

where ω is in radians per second, θ is in radians, and t is in seconds.

TANGENTIAL SPEED AND VELOCITY

For a revolving object, the *tangential speed*, symbolized v and expressed in meters per second (m/s), is the rate at which the object moves around a circular path as if that path were a straight line. When the angular speed is held constant, the tangential speed increases in direct proportion to the radius.

Let ω represent the angular speed of a revolving object, r represent the radius of revolution, θ represent the angular displacement, and t represent elapsed time. Then the tangential speed v can be determined by either of these formulas:

$$v = r\omega$$
$$v = r\,(d\theta\,/\,dt)$$

where v is in meters per second, r is in meters, ω is in radians per second, θ is in radians, and t is in seconds.

Tangential velocity is a vector quantity symbolized by the lowercase bold letter **v** just like linear velocity. In uniform circular motion, dir **v** constantly changes but $|\mathbf{v}|$ or v remains constant. When we talk about a tangential velocity vector for an object following any path that is not a straight line, we must specify it as an instantaneous quantity so the direction can be defined, as is done for a single point in Fig. 3-1A.

CENTRIPETAL ACCELERATION

When an object or point revolves in uniform circular motion, the tangential speed remains constant, but the tangential velocity changes because dir **v** rotates. Any change in velocity represents acceleration. In linear motion, acceleration exhibits itself as a change in speed but not direction. In uniform circular motion, acceleration occurs as a change in direction but not tangential speed, and is called *centripetal acceleration*. This quantity is a vector that can be denoted as $\mathbf{a_c}$. Its magnitude can be represented as $|\mathbf{a_c}|$ or a_c.

Let v represent the tangential speed of a revolving object, r represent the radius of revolution, and ω represent the angular speed. The centripetal acceleration a_c can be determined by either of these formulas:

$$a_c = v^2\,/\,r$$
$$a_c = \omega^2 r$$

where a_c is in meters per second squared, v is in meters per second, r is in meters, and ω is in radians per second.

As an object orbits a central point in a perfect circle at uniform angular speed, the vectors **r**, **v**, and $\mathbf{a_c}$ all rotate at the same constant rate, but they point in different directions. At any instant in time, dir **v** is one-quarter of a circle "ahead" of dir **r**, and dir $\mathbf{a_c}$ is one-quarter of a circle "ahead" of dir **v**. Technically, we write that dir **v** leads dir **r** by $\pi/2$, and dir $\mathbf{a_c}$ leads dir **v** by $\pi/2$. (In physics, the absence of any unit in the depiction of an angle means that the angle is meant to be expressed in radians.) This is shown in Fig. 3-2 for an object in uniform circular motion at the start of an orbit (A), one-quarter of the way around (B), halfway around (C), and three-quarters of the way around (D). The originating points for all the vectors are placed at the origin of a *polar coordinate system* so their relative directions are easy to compare. The vector lengths in these graphs are arbitrary.

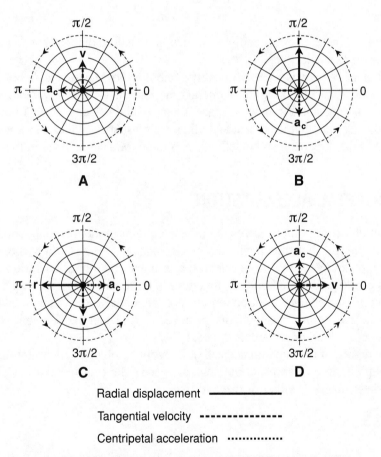

Figure 3-2 Relative directions of the radial displacement vector **r**, the tangential velocity vector **v**, and the centripetal acceleration vector $\mathbf{a_c}$ for an object in uniform circular motion at the start of an orbit (A), one-quarter of the way around (B), halfway around (C), and three-quarters of the way around (D). Vector lengths are arbitrary.

PROBLEM 3-1

Imagine a satellite in a perfectly circular orbit around a planet, such that the satellite is always 7225 km from the center of the planet and takes 121.5 minutes (min) to complete each orbit. Calculate the angular speed, the tangential speed, and the centripetal acceleration for this satellite. Consider the value of π to be 3.14159. Express the answers to four significant figures.

SOLUTION 3-1

First, convert the input data to meters for the radius r and seconds for the period T. Remembering that 1 km = 1000 m and 1 min = 60 s, we find that:

$$r = 7.225 \times 10^6 \text{ m}$$
$$T = 7290 \text{ s}$$

The angular speed ω is therefore:

$$\begin{aligned} \omega &= 2\pi / T \\ &= 2.000 \times 3.14159 / 7290 \\ &= 8.619 \times 10^{-4} \text{ rad/s} \end{aligned}$$

To find the tangential speed, we multiply the angular speed by the radius, as follows:

$$\begin{aligned} v &= r\omega \\ &= 7.225 \times 10^6 \times 8.619 \times 10^{-4} \\ &= 6227 \text{ m/s} \end{aligned}$$

We can find centripetal acceleration in terms of tangential speed and radius, or in terms of angular speed and radius. First, in terms of tangential speed and radius:

$$\begin{aligned} a_c &= v^2 / r \\ &= (6227)^2 / (7.225 \times 10^6) \\ &= (3.878 \times 10^7) / (7.225 \times 10^6) \\ &= 5.367 \text{ m/s}^2 \end{aligned}$$

Second, in terms of angular speed and radius:

$$\begin{aligned} a_c &= \omega^2 r \\ &= (8.619 \times 10^{-4})^2 \times 7.225 \times 10^6 \\ &= 7.429 \times 10^{-7} \times 7.225 \times 10^6 \\ &= 5.367 \text{ m/s}^2 \end{aligned}$$

PROBLEM 3-2

Suppose the satellite in the previous situation has a mass of 100.0 kg. What is the force F associated with the centripetal acceleration?

SOLUTION 3-2

To solve this, remember the formula for force in terms of mass and acceleration. Then:

$$F = ma_c$$
$$= 100.0 \times 5.367$$
$$= 536.7 \text{ N}$$

Rotational Dynamics

A mass "resists" any attempt to make it move or to change its linear speed. Isaac Newton quantified this behavior in his laws of motion. What about getting a solid mass to rotate or change its angular speed? Similar laws apply, but some new quantities must be defined.

TORQUE AS A VECTOR

Torque is a phenomenon that starts an object rotating or that changes the angular speed of an object that is already rotating. Mathematically, torque is a vector quantity symbolized as $\boldsymbol{\tau}$ (the lowercase, bold Greek letter tau). Intuitively, it can be thought of as "twisting action," although that is not a rigorous definition.

Imagine a solid, massive disk set on a frictionless bearing, such as the active element in a perfect gyroscope. Suppose a force **F** is applied at point P at a distance r from the center of the disk, as shown in Fig. 3-3. Also suppose that **F** lies in the same plane as the disk, and subtends an angle θ with respect to the radius vector **r** extending from the center of the disk to point P. Then the torque $\boldsymbol{\tau}$ produced by **F** is equal to the cross product of the radius and force vectors:

$$\boldsymbol{\tau} = \mathbf{r} \times \mathbf{F}$$

The torque vector $\boldsymbol{\tau}$ in the scenario of Fig. 3-3 points directly at you, out of the page, perpendicular to the disk, according to the *right-hand rule for cross products*.

THE RIGHT-HAND RULE

If you've forgotten the right-hand rule for cross products, here's a brief review. Imagine two vectors **r** and **F** that lie in the same plane, such as in Fig. 3-3. Let θ be the angle between dir **r** and dir **F**. Curl the fingers of your right hand and straighten your thumb so it is perpendicular to the planes defined by your fingers, the way you would do if you were gripping a golf club. Position your hand so your

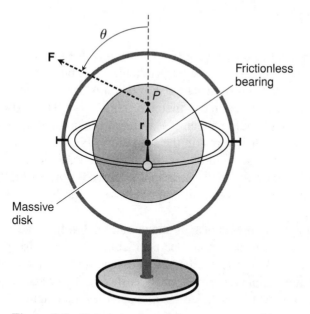

Figure 3-3   Determination of the torque caused by a force applied to a rotatable disk. Illustration for Problems and Solutions 3-3 through 3-6.

fingers curl in the rotational sense of θ (counterclockwise in the case of Fig. 3-3), while the planes containing your fingers are parallel to the plane defined by **r** and **F**. When you do this, your thumb points in the direction of **r** × **F** if sin θ is positive ($0 < \theta < \pi$), and in the opposite direction from **r** × **F** if sin θ is negative ($\pi < \theta < 2\pi$). If sin θ is equal to zero ($\theta = 0$ or $\theta = \pi$), then **r** × **F** is the zero vector. Remember that if **r** × **F** is non-zero, then it is always *exactly* perpendicular to the plane defined by **r** and **F**.

TORQUE MAGNITUDE AS A SCALAR

If we consider only torque magnitude and forget about the rather counterintuitive nature of torque direction, then it becomes a scalar quantity symbolized by τ (the lowercase, italic Greek letter tau) and we can express the relationship among torque magnitude, radial distance, force magnitude, and the angle θ like this:

$$\tau = rF \sin \theta$$

where τ is in newton-meters, r is in meters, F is in newtons, and θ is in whatever unit of angular measure we like (radians preferred).

LEVER ARM

In the situation illustrated by Fig. 3-3, not all of the force **F** contributes to disk rotation. That would be true if **F** were perpendicular to **r**, but that isn't the case here. The fraction of **F** that contributes to disk rotation is equal to the sine of the angle between **F** and **r**. This fraction can therefore range from 0 to 1 (0% to 100%).

Think about what would happen if **F** were applied at point P either straight up or straight down to a disk at rest in the scenario of Fig. 3-3. Then sin θ would be equal to 0. In neither case would **F** contribute to torque; **F** would simply stress the disk. At the other extreme, if **F** were applied directly toward the left or directly toward the right, then it would all contribute to torque because sin θ would be either 1 or −1.

Consider the *line of action* of **F** as shown in Fig. 3-4. Imagine another line running through the line of action and perpendicular to it, and also passing through the center of the disk. Let **r**$_L$ be the vector extending from the center of the disk to the intersection point of the two lines, as shown. This vector is known as the *lever arm*. In some texts it is called the *moment arm*. Its magnitude r_L is:

$$r_L = r \sin \theta$$

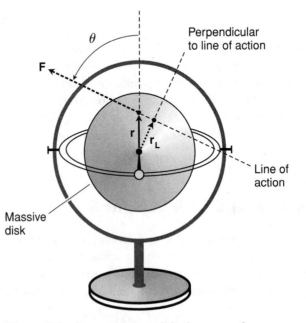

Figure 3-4 Determination of the lever arm for a force applied to a rotatable disk.

Note that $\mathbf{r_L}$ is perpendicular to $\mathbf{F}$, and $\mathbf{r_L}$ subtends an angle of $\pi/2 - \theta$ with respect to $\mathbf{r}$.

The concept of the lever arm allows us to express the torque magnitude as a simple product:

$$\tau = r_L F$$

where τ is in newton-meters, r_L is in meters, and F is in newtons.

ANGULAR AND TANGENTIAL ACCELERATION

In uniform circular motion, the angular speed remains constant. This produces centripetal acceleration, but the *angular acceleration* is zero. Torque is associated with a change in angular speed, so the angular acceleration is non-zero.

Angular acceleration occurs when a non-rotating object begins to rotate. It can also appear as an increase or a decrease in the angular speed for an object that is already rotating. Angular acceleration magnitude can be denoted by α (the lowercase, italic Greek letter alpha) and is the derivative of the angular speed with respect to time:

$$\alpha = d\omega / dt$$

When the angular speed of a rotating or revolving point changes, the tangential speed changes too. Any change in tangential speed is known as *tangential acceleration*. Let's symbolize this as a_t to contrast it with centripetal acceleration, a_c. Tangential acceleration, also called *linear acceleration*, is equal to the angular acceleration multiplied by the radius:

$$a_t = r\alpha$$

where a_t is in meters per second squared, r is in meters, and α is in radians per second squared. (Note that when combining units in formulas like this, radians vanish. That's because the radian is theoretically without dimension.)

Tangential acceleration, like angular acceleration, is often expressed as an instantaneous quantity. The tangential acceleration vector $\mathbf{a_t}$, unlike the centripetal acceleration vector $\mathbf{a_c}$, points in the same direction as the tangential velocity vector $\mathbf{v}$.

ROTATIONAL INERTIA

Torque can be thought of as the rotational equivalent of straight-line force. Angular acceleration can be thought of as the rotational equivalent of straight-line acceleration. There's also a rotational equivalent of mass: *rotational inertia*, symbolized by the uppercase italic letter I. In some texts it is called *moment of inertia*.

Once again, think of a rotating disk in a perfect gyroscope. Let **F** be a force applied at a location P on the disk, but this time, suppose **F** is perpendicular to **r** as shown in Fig. 3-5. Now imagine that P is not quite a perfect geometric point, but an arbitrarily tiny portion of the disk that we call a *point mass*. Let m be the mass of portion P. Let I_p be the rotational inertia of P. This quantity is equal to the mass of P times the square of the radial distance r of portion P from the center of the disk:

$$I_p = mr^2$$

where I is in kilogram meters squared (kg · m²), m is in kilograms, and r is in meters.

The rotational inertia of a solid disk having total mass M, radius R measured from the center to the edge, and uniform density throughout can be found by this formula:

$$I = MR^2 / 2$$

where I is in kilogram meters squared, M is in kilograms, and R is in meters.

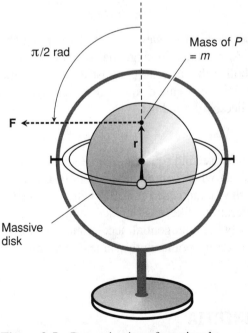

Figure 3-5 Determination of rotational inertia when a tangential force is applied to a rotatable disk.

NEWTON'S SECOND LAW FOR ROTATIONAL MOTION

When the rotational inertia of a solid object is known, the torque required to produce a given angular acceleration can be found as follows:

$$\tau = I\alpha$$

where τ is the torque in newton-meters, I is the rotational inertia in kilogram
meters squared, and α is the angular acceleration in radians per second squared.
You will recognize this as the rotational analog of Newton's second law for
linear motion:

$$F = ma$$

PROBLEM 3-3
In the scenario of Fig. 3-3, suppose that the radius of the disk is 500 mm, the distance of P from the center is 350 mm, the force applied to P is 3.30 N in the
direction shown, and the angle between $\mathbf{r}$ and $\mathbf{F}$ is 65.0°. Solve for τ and express
it to three significant figures.

SOLUTION 3-3
In order to find the magnitude of the torque vector, recall this formula:

$$\tau = rF \sin \theta$$

In this situation, $r = 0.350$ m, $F = 3.30$ N, and $\theta = 65.0°$. Plugging in the numbers,
we obtain:

$$\begin{aligned}
\tau &= 0.350 \times 3.30 \times \sin 65.0° \\
&= 0.350 \times 3.30 \times 0.906 \\
&= 1.05 \text{ N} \cdot \text{m}
\end{aligned}$$

PROBLEM 3-4
What is dir τ in the above described situation?

SOLUTION 3-4
To figure out the direction in which τ points, recall that the torque vector is equal
to the cross product of the radius and force vectors:

$$\boldsymbol{\tau} = \mathbf{r} \times \mathbf{F}$$

Because the angle θ is expressed counterclockwise and its sine is positive, the
right-hand rule for cross products tells us that $\mathbf{r} \times \mathbf{F}$ points towards us, straight out
of the page, perpendicular to the plane of the disk.

PROBLEM 3-5

Suppose the mass of the entire disk in the above situation is 24.0 kg. What is the rotational inertia of the disk? Express the answer to three significant figures.

SOLUTION 3-5

Note that the overall radius of the disk, R, is 500 mm or 0.500 m. Let $M = 24.0$. Then:

$$I = MR^2 / 2$$
$$= 24.0 \times (0.500)^2 / 2.00$$
$$= 24.0 \times 0.250 / 2.00$$
$$= 3.00 \text{ kg} \cdot \text{m}^2$$

PROBLEM 3-6

In the above described situation, what is the angular acceleration of the disk? If it starts from a non-rotating state, how fast will it be spinning after 1.00 s? After 2.00 s? Express the answers to three significant figures.

SOLUTION 3-6

To solve this, use the following formula and then plug in values already calculated:

$$\tau = I\alpha$$
$$1.05 = 3.00 \times \alpha$$
$$\alpha = 1.05 / 3.00$$
$$= 0.350 \text{ rad/s}^2$$

If it starts out with angular velocity of zero (not rotating at all), then after 1.00 s it will be spinning at 0.350 rad/s, and after 2.00 s it will be spinning at 0.700 rad/s.

Angular Momentum and Kinetic Energy

An object moving in a straight line possesses linear momentum and kinetic energy that can be quantified. For rotating or revolving objects, the counterparts are *angular momentum* and *rotational kinetic energy*.

LINEAR VS. TANGENTIAL MOMENTUM

Let's review the formulas for linear momentum. Given an object with mass m and velocity vector **v** moving in a straight line, the linear momentum vector **p** is:

$$\mathbf{p} = m\mathbf{v}$$

and dir **p** = dir **v**. The scalar relation among linear momentum magnitude p, mass m, and linear speed v is:

$$p = mv$$

where p is in kilogram meters per second, m is in kilograms, and v is in meters per second.

Now consider a point mass P that revolves in a perfect circle around a central point at a constant angular speed, as shown in Fig. 3-6. Let m be the mass of P, and let **v** be its tangential velocity vector at a given instant in time. At that instant (or at any other instant), the *tangential momentum* vector **p** of the revolving point mass P is identical to its linear momentum vector. If the angular speed of P is constant over time, so is its tangential speed v, and so is its tangential momentum magnitude p.

Note that dir **p** changes over time, rotating at the same angular speed and in the same sense as P revolves, even though the vector magnitude p remains constant.

ANGULAR MOMENTUM FOR A REVOLVING OBJECT

Angular momentum is the circular-motion analog of linear momentum. This is technically a vector quantity, symbolized by the uppercase, bold letter **L**. Angular momentum magnitude is symbolized by the uppercase italic L.

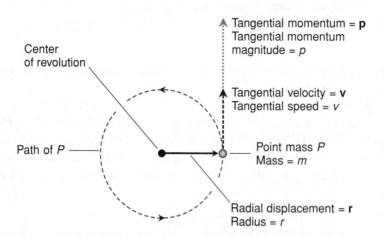

Figure 3-6 An object of mass m revolving at constant angular speed, showing radial displacement **r**, radius r, tangential velocity **v**, tangential speed v, tangential momentum **p**, and tangential momentum magnitude p.

Given the situation shown in Fig. 3-6, the angular momentum magnitude can be found by multiplying the tangential momentum magnitude by the radius:

$$L = rp = rmv$$

where L is in kilogram meters squared per second (kg · m²/s), r is in meters, p is in kilogram meters per second, m is in kilograms, and v is in meters per second. Remembering that $v = r\omega$ where ω is the angular speed, we can express the angular momentum as follows:

$$L = rm \, (r\omega) = r^2 m\omega$$

where L is in kilogram meters squared per second (kg · m²/s), r is in meters, m is in kilograms, and ω is in radians per second.

In terms of vectors, we express the angular momentum of a revolving point mass as follows:

$$\mathbf{L} = \mathbf{r} \times \mathbf{p}$$

The right-hand rule for cross products tells us that in the scenario of Fig. 3-6, dir **L** is out of the page toward us, exactly perpendicular to the plane defined by vectors **r** and **p**.

ANGULAR MOMENTUM FOR A ROTATING SOLID OBJECT

Now imagine a rotating solid disk with rotational inertia I and angular speed ω. In this case, the angular momentum magnitude is the product of the rotational inertia and the angular speed:

$$L = I\omega$$

where L is in kilogram meters squared per second, I is in kilogram meters squared, and ω is in radians per second.

A solid rotating disk can be thought of as a large number of tiny pieces glued together, each of which revolves around the center in the same sense (counterclockwise or clockwise) as the rotation of the disk. Let's call any arbitrary one of these pieces S, as shown in Fig. 3-7. Every S has its own angular momentum vector $\mathbf{L_s}$ that is the cross product of its radius vector $\mathbf{r_s}$ and its tangential momentum vector $\mathbf{p_s}$. That is:

$$\mathbf{L_s} = \mathbf{r_s} \times \mathbf{p_s}$$

Because every S revolves in the same sense that the disk rotates, and because the disk is solid, the vector **L** for the whole disk is equal to the vector sum of the $\mathbf{L_s}$ vectors for all the tiny portions S that make up the disk. It follows that for any S, dir **L** is the same as dir $\mathbf{L_s}$ (although the magnitude of **L** is much greater than the

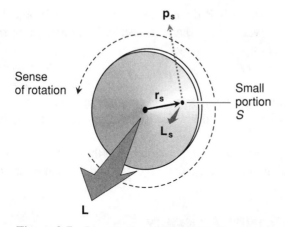

Figure 3-7   In a rotating disk, dir **L** is the same as dir $\mathbf{L}_s$ for any arbitrary portion S. Here, $\mathbf{r}_s$ is the radius vector of S, $\mathbf{p}_s$ is the tangential momentum of S, and $\mathbf{L}_s = \mathbf{r}_s \times \mathbf{p}_s$.

magnitude of $\mathbf{L}_s$). This rule for dir **L** can be applied to any solid object, not only a disk. Here is the general principle stated in two parts:

- If you see a solid object rotating counterclockwise, **L** points generally at you, perpendicular to the plane of rotation
- If you see a solid object rotating clockwise, **L** points generally away from you, perpendicular to the plane of rotation

CONSERVATION OF ANGULAR MOMENTUM

Recall the rule for conservation of linear momentum stated in Chapter 1. According to that principle, the momentum vector of an object moving in a straight line remains constant if and only if the net force acting on that object is always zero. If **F** is the net force vector, **0** (in boldface) represents the zero vector, **p** is the linear momentum vector, and **k** is a constant vector, then we can write this logical formula:

$$(\mathbf{F} = \mathbf{0}) \Leftrightarrow (\mathbf{p} = \mathbf{k})$$

where the $\Leftrightarrow$ symbol means "if and only if." The statement on the left side of the $\Leftrightarrow$ symbol is *logically equivalent* to the statement on its right side.

A similar rule applies when we substitute torque for force and angular momentum for linear momentum. The angular momentum vector of a rotating solid object remains constant if and only if the net torque acting on that object is always zero. If τ is the net torque vector, $\mathbf{0}$ is the zero vector, $\mathbf{L}$ is the angular momentum vector, and $\mathbf{k}$ is a constant vector, then:

$$(\tau = \mathbf{0}) \Leftrightarrow (\mathbf{L} = \mathbf{k})$$

This is a statement of the *law of conservation of angular momentum*. It arises directly from Newton's second law for linear motion and its analog for rotational motion.

When the net torque vector on a solid object (the vector sum of all the torques, if there are more than one) is zero, that object is said to be in *rotational equilibrium*.

ROTATIONAL KINETIC ENERGY

An object moving in a straight line at a constant speed has kinetic energy. Likewise, an object rotating or revolving at a constant rate also has kinetic energy. *Rotational kinetic energy*, like its linear-motion counterpart, is a scalar quantity. Let's denote rotational kinetic energy as E_{rk}.

Imagine a point mass P having mass m revolving in a perfect circle at constant tangential speed v around a central point, as shown in Fig. 3-8. The rotational kinetic energy $E_{kr\text{-}p}$ of this point mass can be found using the formula for linear-motion kinetic energy. That is:

$$E_{kr\text{-}p} = mv^2 / 2$$

where $E_{kr\text{-}p}$ is in joules, m is in kilograms, and v is in meters per second.

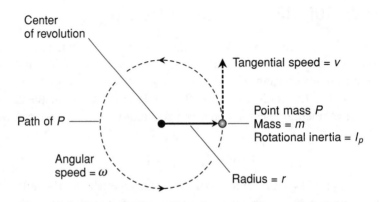

Figure 3-8 Determination of kinetic energy for a revolving object P of mass m revolving at a constant rate, showing radius r, tangential speed v, angular speed ω, and rotational inertia I_p.

What if we are given the rotational inertia I_p and the angular speed ω of P, rather than the mass m and the tangential speed v? How can we determine the kinetic energy of P in that case? To figure this out, we recall the formula for rotational inertia of a revolving point mass in terms of mass and radius, along with the formula for tangential speed in terms of radius and angular speed. These formulas appeared earlier in this chapter:

$$I_p = mr^2$$
$$v = r\omega$$

where I_p is in kilogram meters squared, m is in kilograms, r is in meters, v is in meters per second, and ω is in radians per second. We can rearrange the first of these formulas so it gives us mass in terms of rotational inertia and radius. Divide both sides by r^2, getting:

$$m = I_p / r^2$$

Then by substitution:

$$
\begin{aligned}
E_{kr\text{-}p} &= mv^2 / 2 \\
&= (I_p / r^2)(r\omega)^2 / 2 \\
&= (I_p / r^2) r^2 \omega^2 / 2 \\
&= I_p \omega^2 / 2
\end{aligned}
$$

where $E_{kr\text{-}p}$ is in joules, I_p is in kilogram meters squared, and ω is in radians per second. If you want to get picayune here, we can call $E_{kr\text{-}p}$ "revolutional kinetic energy" rather than rotational kinetic energy, because P is revolving, not rotating!

Now suppose that we have a solid rotating disk of uniform density, rather than a single revolving point mass. The disk can be considered as a huge assembly of tiny point masses P, whose overall mass is equal to the sum of all the tiny point masses. Each of the little point masses revolves at the same angular speed ω as the disk rotates. Therefore, the rotational kinetic energy E_{kr} of the disk is equal to the sum of all the little rotational kinetic energies $E_{kr\text{-}p}$. The following formula gives us the rotational kinetic energy of the whole disk:

$$E_{kr} = I\omega^2 / 2$$

where E_{kr} is in joules, I is in kilogram meters squared, and ω is in radians per second.

PROBLEM 3-7
A solid disk has a radius of 220 mm and a mass of 454 g. The disk rotates clockwise at a constant rate, such that the tangential speed of any point on its edge is 45.7 m/s. What is the rotational inertia of the disk? What is the magnitude of its

angular momentum vector? Find these values to three significant figures. What is the direction of the angular momentum vector?

SOLUTION 3-7

First, we must find the angular speed of the disk. We know that for any point on the edge, its distance from the center r is 220 mm or 0.220 m, and its tangential speed v is 45.7 m/s. Recall the formula that relates tangential speed, radius, and angular speed:

$$v = r\omega$$

Therefore:

$$\omega = v / r$$
$$= 45.7 / 0.220$$
$$= 208 \text{ rad/s}$$

Next, we determine the rotational inertia I for the disk. We are told that the radius R of the disk is 0.220 m. The mass M of the disk is 454 g or 0.454 kg. Therefore:

$$I = MR^2 / 2$$
$$= 0.454 \times (0.220)^2 / 2$$
$$= 0.454 \times 0.0484 / 2$$
$$= 0.0110 \text{ kg} \cdot \text{m}^2$$

In order to determine the angular momentum magnitude, we use this formula:

$$L = I\omega$$
$$= 0.0110 \times 208$$
$$= 2.29 \text{ kg} \cdot \text{m}^2/\text{s}$$

According to the rules for the direction of the angular momentum vector **L** outlined above, **L** points perpendicular to the plane of rotation, and away from an observer who sees the disk rotating clockwise.

PROBLEM 3-8

Suppose that in the rotating-disk scenario of the previous problem, a *braking torque* is applied that causes the angular speed to decrease at a constant rate of 20.8 rad/s^2. Determine the magnitude of this torque vector to three significant figures. In what direction does it point? What would happen if the torque were applied in this manner indefinitely?

SOLUTION 3-8

Let's answer the second question first. The applied torque vector τ acts against the angular momentum vector **L**. As the disk slows down, its angular momentum magnitude decreases. That means dir $\boldsymbol{\tau}$ is contrary to dir $\mathbf{L}$, so $\boldsymbol{\tau}$ points generally toward an observer who sees the disk rotating clockwise at a diminishing angular speed, and $\boldsymbol{\tau}$ is perpendicular to the plane of rotation.

We can find the torque magnitude τ based on the rotational inertia I and the angular acceleration α as follows:

$$
\begin{aligned}
\tau &= I\alpha \\
&= 0.0110 \times 20.8 \\
&= 0.229 \text{ N} \cdot \text{m}
\end{aligned}
$$

This is a positive value because we have defined dir $\boldsymbol{\tau}$ as contrary to dir $\mathbf{L}$.

If we were to keep applying $\boldsymbol{\tau}$ in a constant manner, the magnitude of $\mathbf{L}$ would decrease at a constant rate. If the torque were applied for long enough, the disk would eventually come to a halt at a point in time, and then it would start rotating counterclockwise. At the instant the disk was stopped, $\mathbf{L}$ would vanish. Then $\mathbf{L}$ would reappear and grow in magnitude but in the opposite direction, generally toward the observer and in the same direction as $\boldsymbol{\tau}$. After the rotation reversal, the torque would no longer be a braking influence, but would cause positive angular acceleration instead.

The Circular-Motion Model for Oscillation

When a point revolving in a circular path at constant angular speed is observed from an arbitrarily great distance within the plane defined by its revolution, its apparent back-and-forth lateral movement constitutes a regular oscillation pattern known as *simple harmonic motion* (SHM). This is the *circular-motion model for oscillation*.

THE SINUSOID

Once again, imagine that you whirl a ball at the end of a non-elastic string so it describes a perfectly circular path at a constant angular speed in a horizontal plane. As seen from directly above, the ball revolves counterclockwise as shown in Fig. 3-9A. Imagine that you stand in the middle of a huge, flat prairie so people can see you from great distances (and, perhaps, make fun of you out of earshot).

Now suppose that your best friend stands a practically infinite distance away. In the real world, several hundred times the radius of the ball's orbit is good enough. His or her eyes are precisely in the plane defined by the orbit, so there is no evidence of circular motion at all. Your friend looks through a telescope and

Advanced Physics Demystified

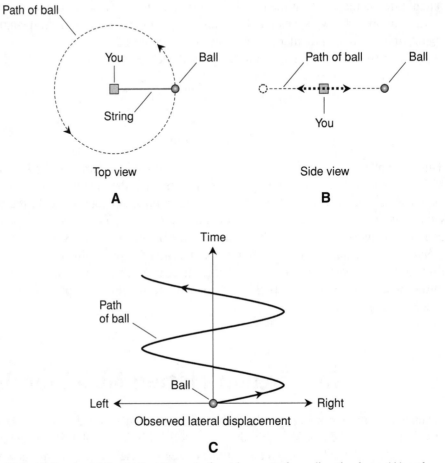

Figure 3-9 A whirling ball at the end of a string, seen from directly above (A) and from directly edgewise (B). At C, a graph of lateral displacement vs. time is shown.

sees the ball oscillating back and forth along a straight, horizontal line (Fig. 9-3B). If your friend graphs the apparent left-to-right movement of the ball against time (Fig. 9-3C), the result is a *sinusoid*, also called a *sine wave* because its shape can be defined on the basis of the trigonometric *sine function*.

When SHM is graphed as a function of displacement vs. time, the result is always a wave of this general shape. Some sinusoids are "taller" or "shorter" than others; some are more or less "stretched out" than others. Some are shifted to the left or the right with respect to others. However, the shape can always be defined in terms of the sine function.

AMPLITUDE

Let θ be the instantaneous angular displacement (in radians) of the ball counterclockwise relative to a defined starting point such that you, the ball, and your friend are all on the same straight line, with the ball in the middle. As the ball revolves, let q be the instantaneous lateral displacement from the center (in meters) as observed by your friend. Then:

$$q = q_{pk+} \sin \theta$$

where q_{pk+} is the maximum positive lateral displacement (in meters), also known as the *positive peak amplitude*. In this example, the positive lateral direction is defined as being toward the right. Figure 3-10 is a graph of this equation for one complete cycle. We can also say this:

$$q = -q_{pk-} \sin \theta$$

where q_{pk-} is the maximum negative lateral displacement (in meters), also called the *negative peak amplitude*. In this scenario, the negative lateral direction is defined as being toward the left. In SHM based on the circular-motion model, the positive and negative peak amplitudes are always equal but opposite:

$$q_{pk+} = -q_{pk-}$$

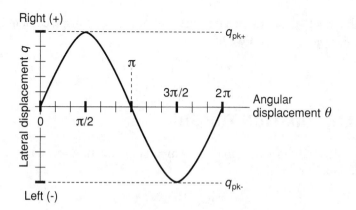

Figure 3-10 Lateral displacement of whirling ball (vertical axis) as a function of angular displacement in radians (horizontal axis).

ANGULAR FREQUENCY

Angular displacement is related to time according to the angular speed of a revolving object. Angular speed is called *angular frequency* when it relates to wave motions or SHM. If ω is the angular frequency and t is the elapsed time as measured from any instant where $q = 0$ and is increasing positively (in Fig. 3-9B, that would be when the ball appears centered and is moving toward the right), then:

$$\theta = \omega t$$

$$q = q_{pk+} \sin \omega t$$

where θ is in radians, ω is in radians per second, and t is in seconds.

PHASE

Phase, also called *phase angle*, is the angular displacement at which we define circular motion, a wave cycle, or SHM as first starting up. It is symbolized by ϕ (the lowercase italic Greek letter phi) and is expressed in radians.

In the preceding pair of equations, $\phi = 0$ because $\theta = 0$ when $t = 0$. But suppose we define the starting instant $t = 0$ as taking place when the ball is at its maximum positive displacement q_{pk+} where $\theta = \pi/2$. In that case, we have $\phi = \pi/2$, and the above equation becomes:

$$q = q_{pk+} \sin (\omega t + \pi/2)$$

If we define $t = 0$ as taking place at maximum negative displacement q_{pk-} where $\theta = 3\pi/2$, then $\phi = 3\pi/2$ and we have this:

$$q = q_{pk+} \sin (\omega t + 3\pi/2)$$

COMPLETE EQUATION FOR SHM

We are finally ready to write an all-encompassing equation for SHM based on the circular-motion model:

$$q = q_{pk+} \sin (\omega t + \phi)$$

where q and q_{pk+} are in meters, ω is in radians per second, t is in seconds, and ϕ is in radians. By convention, ϕ is expressed as a non-negative value less than 2π.

If the sense of revolution in the circular-motion model is changed from counterclockwise to clockwise or vice-versa but the starting point remains the same, we have *phase reversal*. In SHM represented by a sinusoid, that's the equivalent of shifting the phase by half a cycle without reversing the sense of revolution, thereby increasing or decreasing the value of ϕ by exactly π rad.

PERIOD AND CYCLIC FREQUENCY

We have now defined amplitude, angular frequency, and phase for SHM. Two more parameters are sometimes specified as well: the *period* and the *cyclic frequency* (more often called simply the *frequency*).

The period of rotation, revolution, a wave, or SHM is the length of time for one complete orbit or cycle to take place. It is a scalar quantity symbolized by the italic uppercase letter T, and is expressed in seconds. Once in awhile, the period is expressed in units other than seconds, but it should always be reduced to seconds before calculations are made.

The cyclic frequency of rotation, revolution, a wave, or SHM is the number of complete orbits or cycles that occur in 1 s. This quantity is a scalar symbolized by the italic lowercase letter f. The standard unit of frequency is the *hertz* (Hz), which represents one cycle per second (1 cps). Often, you will see larger units such as the *kilohertz* (kHz), *megahertz* (MHz), *gigahertz* (GHz), and *terahertz* (THz). Here are the comparisons:

$$1 \text{ kHz} = 1000 \text{ Hz}$$
$$1 \text{ MHz} = 1000 \text{ kHz} = 10^6 \text{ Hz}$$
$$1 \text{ GHz} = 1000 \text{ MHz} = 10^9 \text{ Hz}$$
$$1 \text{ THz} = 1000 \text{ GHz} = 10^{12} \text{ Hz}$$

Frequency values should always be reduced to hertz before calculations are made.

The period T, the cyclic frequency f, and the angular frequency ω in SHM are related as follows:

$$T = 1 / f = 2\pi / \omega$$
$$f = \omega / (2\pi) = 1 / T$$
$$\omega = 2\pi / T = 2\pi f$$

where T is in seconds, f is in hertz, and ω is in radians per second.

The complete equation for SHM, stated in the previous section, can also be written in either of the following two ways by simple substitution for the angular frequency:

$$q = q_{\text{pk+}} \sin \left[(2\pi t / T) + \phi \right]$$
$$q = q_{\text{pk+}} \sin (2\pi f t + \phi)$$

An Oscillating Spring

When an *ideal spring* is repeatedly flexed by an object attached to one of its ends, the motion of that object constitutes SHM. A spring stores potential energy when compressed or stretched, and then releases it as kinetic energy when it is let go. An ideal spring has zero mass, and it stores and releases energy with 100-percent efficiency. Figure 3-11 shows an example in which a rolling cart with frictionless wheels is attached, by means of an ideal spring, to an anchored-down platform.

FORCE VS. DISPLACEMENT

If the cart is set at the point where the spring is under no tension, the lateral force magnitude F that the spring exerts on the cart is zero. Let's assign this point the displacement magnitude $q = 0$. If we push the cart toward the right (positive displacement), the spring exerts a force toward the left, in the negative direction. If we pull the cart toward the left (negative displacement), the spring exerts a force toward the right, in the positive direction.

Now imagine what happens if we push the cart to the right, compressing the spring until the displacement is q_{pk+} as indicated on the scale in Fig. 3-11, and then

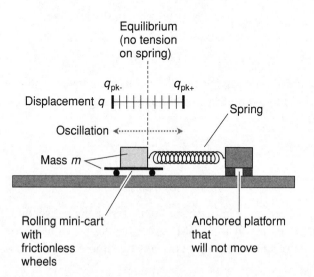

Figure 3-11 A spring attached to a rolling cart of mass m with frictionless wheels. The cart rolls back and forth between extreme displacement points q_{pk-} (left) and q_{pk+} (right).

let the cart go. It oscillates between the extremes q_{pk+} and q_{pk-} at a constant angular frequency. Because there is no friction between the cart and the air or the surface on which it rolls, no friction in the wheels, and no energy loss in the spring, the oscillation continues in this way indefinitely, unless or until an outside force intervenes.

HOOKE'S LAW

Two factors determine the angular frequency at which an ideal spring-loaded system of the sort shown in Fig. 3-11 will oscillate: the *spring constant* and the mass of the object attached to the movable end of the spring. Remember that an ideal spring has no mass of its own, so the mass of the spring does not figure into this theoretical situation.

The spring constant expresses the force exerted by a spring as a function of the displacement (stretching or compression) from the equilibrium position. It is given in newtons per meter (N/m or N · m^{-1}). Every spring has a spring constant that depends on what it's made of and how it's constructed. It never changes unless the spring is permanently deformed by excessive stretching. Let's call this constant k_s. In an ideal spring-loaded system, the force magnitude F and the positive peak amplitude q_{pk+} are related as follows:

$$F = -k_s\, q_{pk+}$$

where F is in newtons, k_s is in newtons per meter, and q_{pk+} is in meters. The minus sign in the right-hand side of this equation reflects the fact that the direction of the force vector $\mathbf{F}$ opposes the direction of the displacement vector $\mathbf{q}$. This equation is known as *Hooke's law for ideal springs*.

Let ω be the angular frequency of oscillation in the system of Fig. 3-11. Let m be the mass of the movable cart including its payload. Then:

$$\omega = (k_s\,/\,m)^{1/2}$$

where ω is in radians per second, k_s is in newtons per meter, and m is in kilograms. From this, and from the known relationship among angular frequency, cyclic frequency, and period, we can derive these formulas:

$$f = (1\,/\,2\pi)\,(k_s\,/\,m)^{1/2}$$
$$T = 2\pi\,(m\,/\,k_s)^{1/2}$$

where f is in hertz, k_s is in newtons per meter, m is in kilograms, and T is in seconds. The cyclic frequency f for a particular spring-loaded system is called the *natural frequency* or *resonant frequency*.

ENERGY STORAGE AND RELEASE

Imagine that we push or pull the cart in the situation of Fig. 3-11 so the spring is flexed, and then let the cart go so it oscillates. When the cart is at either extreme where $q = q_{pk-}$ or $q = q_{pk+}$, its instantaneous linear speed is zero, the force magnitude F exerted by the spring is maximum either positively (toward the right) or negatively (toward the left), and all the system energy is potential. When the cart is centered so that $q = 0$, its instantaneous linear speed is maximum, the force exerted by the spring is zero, and all the system energy is kinetic. At intermediate points where $q_{pk-} < q < 0$ or $0 < q < q_{pk+}$, the force magnitude F is somewhere between its positive maximum and its negative maximum, some of the system energy is potential, and some of the system energy is kinetic.

From the previous chapter, remember the formula for mechanical (total) energy E_m in terms of potential energy E_p and kinetic energy E_k in a system. Here it is again:

$$E_m = E_p + E_k$$

In a system where SHM is taking place, E_m is constant, provided there is no friction or other energy loss. This is *conservation of mechanical energy*. In the scenario of Fig. 3-11, it means that although the energy keeps changing form, and although the ratio of potential to kinetic energy is always fluctuating, the total system energy never varies. The system keeps "robbing the cart and payload to pay the spring" and vice versa, endlessly.

POTENTIAL AND KINETIC ENERGY

When a spring is compressed or stretched, the instantaneous potential energy E_p it stores is proportional to the spring constant k_s, and is also proportional to the square of the instantaneous displacement magnitude q, as follows:

$$E_p = k_s q^2 / 2$$

where E_p is in joules, k_s is in newtons per meter, and q is in meters.

If the movable cart in the situation of Fig. 3-11 is pushed to the right until its displacement magnitude is $+q_0$, a certain amount of potential energy E_{p0} is stored in the spring. If the cart is then released, it begins to move, passing through the equilibrium point where it attains its highest speed. At this point, the kinetic energy in the cart and payload, E_{k0}, is the same as E_{p0} was when the spring was fully compressed, and the potential energy in the spring has dropped to zero. After it passes the equilibrium point, the cart slows down but continues to displacement magnitude $-q_0$, where it comes to rest for an instant before reversing its direction. At this extreme point, the potential energy in the spring is E_{p0} once again, and the kinetic energy in the cart and payload has dropped to zero. After that, the cart reverses direction, and the cycle continues.

In a situation such as that shown in Fig. 3-11, the kinetic energy E_k of the cart and payload is directly proportional to its total mass m, and is also proportional to the square of its instantaneous linear speed v, as follows:

$$E_k = m \, v^2 / 2$$

where E_k is in joules, m is in kilograms, and v is in meters per second. You will recognize this as the formula for kinetic energy in linear motion.

Given the above formulas for E_p and E_k, along with the fact that their sum is always equal to E_m, we can derive a formula for the instantaneous linear speed v of the cart in terms of E_m, k_s, q, and m. Here it is:

$$v = [(2E_m - k_s \, q^2) / m]^{1/2}$$

where v is in meters per second, E_m is in joules, k_s is in newtons per meter, q is in meters, and m is in kilograms.

PROBLEM 3-9

Derive a formula for the instantaneous linear speed v of the cart in Fig. 3-11 as a function of the instantaneous displacement q, the spring constant k_s, the mass m, and the positive peak displacement q_{pk+}.

SOLUTION 3-9

When the spring is at its positive peak displacement, all the energy in the system is potential and none of it is kinetic, so $E_m = E_p$. This fact, along with the fact that E_m is constant throughout the oscillation cycle, lets us find E_m in terms of k_s and q_{pk+}, as follows:

$$E_m = k_s \, q_{pk+}^2 / 2$$

where E_m is in joules, k_s is in newtons per meter, and q_{pk+} is in meters. By substitution in the above formula for v, we obtain:

$$v = \{[2 \, (k_s \, q_{pk+}^2 / 2) - k_s \, q^2] / m\}^{1/2}$$

where v is in meters per second, k_s is in newtons per meter, q_{pk+} and q are in meters, and m is in kilograms.

PROBLEM 3-10

In the scenario shown by Fig. 3-11, suppose the mass of the cart and payload is 400 g, the spring constant is 1.60 N/m, the spring is ideal, and the system has no friction. The cart is initially at rest at the equilibrium point of the spring. Then the cart is pushed to the right against the spring for a distance of 300 mm and let go. The cart oscillates back and forth. To three significant figures, find:

1. The angular frequency of oscillation
2. The cyclic frequency of oscillation
3. The period of oscillation
4. The maximum potential energy held by the spring
5. The maximum kinetic energy in the cart and payload
6. The mechanical energy of the system
7. The instantaneous speed of the cart at the point in the cycle where the instantaneous displacement is −200 mm
8. The kinetic energy in the cart and payload at the point in the cycle where the instantaneous displacement is −200 mm
9. The potential energy held by the spring at the point in the cycle where the instantaneous displacement is −200 mm

SOLUTION 3-10

1. We are given k_s = 1.60 N/m and m = 400 g = 0.400 kg. Therefore:

$$\omega = (k_s / m)^{1/2}$$
$$= (1.60 / 0.400)^{1/2}$$
$$= 4^{1/2}$$
$$= 2.00 \text{ rad/s}$$

2. The cyclic frequency f in hertz is equal to the angular frequency ω divided by 2π. If we consider the value of π to be 3.14, then:

$$f = \omega / (2\pi)$$
$$= 2.00 / (2.00 \times 3.14)$$
$$= 2.00 / 6.28$$
$$= 0.318 \text{ Hz}$$

3. The period T of oscillation in seconds is equal to 2π divided by the angular frequency ω. Therefore:

$$T = 2\pi / \omega$$
$$= (2.00 \times 3.14) / 2.00$$
$$= 3.14 \text{ s}$$

4. We are told that the maximum displacement q_{pk+} is equal to 300 mm. This is 0.300 m. Therefore, the maximum potential energy $E_{p\text{-max}}$ held by the spring is:

$$E_{p\text{-max}} = k_s \, q_{pk+}{}^2 / 2$$
$$= 1.60 \times (0.300)^2 / 2.00$$
$$= 1.60 \times 0.090 / 2.00$$
$$= 0.0720 \text{ J}$$

If you like, you can call this 72.0 mJ, where 1 mJ = 0.001 J.

5. The maximum kinetic energy $E_{k\text{-max}}$ of the cart and its payload is manifest at the instant when there is no potential energy in the spring, because all of its potential energy has been converted to kinetic energy. Therefore:

$$E_{k\text{-max}} = E_{p\text{-max}}$$
$$= 0.0720 \text{ J}$$
$$= 72.0 \text{ mJ}$$

6. The mechanical energy E_m in the system is always equal to the sum of the potential energy held by the spring and the kinetic energy of the cart and its payload. We know that when $E_p = 0$, $E_k = E_{k\text{-max}}$, and also that when $E_k = 0$, $E_p = E_{p\text{-max}}$. Therefore:

$$E_m = E_{p\text{-max}} = E_{k\text{-max}}$$
$$= 0.0720 \text{ J}$$
$$= 72.0 \text{ mJ}$$

7. We know, from previous calculations and from data we are given, the following:

$$E_m = 0.0720 \text{ J}$$
$$k_s = 1.60 \text{ N/m}$$
$$q = -200 \text{ mm} = -0.200 \text{ m}$$
$$m = 400 \text{ g} = 0.400 \text{ kg}$$

Finding the instantaneous linear speed is just a matter of plugging numbers into the appropriate formula and grinding out the result:

$$\begin{aligned}
v &= [(2E_m - k_s q^2) / m]^{1/2} \\
&= \{[2.000 \times 0.0720 - 1.60 \times (-0.200)^2] / 0.400\}^{1/2} \\
&= [(0.144 - 0.0640) / 0.400]^{1/2} \\
&= (0.080 / 0.400)^{1/2} \\
&= (0.200)^{1/2} \\
&= 0.447 \text{ m/s}
\end{aligned}$$

8. We know the mass m of the cart and payload, and we also know its instantaneous linear speed v when the instantaneous displacement is $q = -0.200$ m. According to the formula for kinetic energy in terms of these parameters:

$$\begin{aligned}
E_k &= m v^2 / 2 \\
&= 0.400 \times (0.447)^2 / 2 \\
&= 0.400 \times 0.200 / 2 \\
&= 0.0400 \text{ J} \\
&= 40.0 \text{ mJ}
\end{aligned}$$

9. The potential energy E_p held by the spring at any given instant is equal to the difference between the system mechanical energy E_m and the kinetic energy E_k of the cart and payload:

$$E_p = E_m - E_k$$
$$= 0.0720 - 0.040$$
$$= 0.0320 \text{ J}$$
$$= 32.0 \text{ mJ}$$

Quiz

This is an "open book" quiz. You may refer to the text in this chapter. A good score is 8 correct. Answers are in the back of the book.

1. Suppose that in the scenario shown by Fig. 3-9A, the ball revolves clockwise instead of counterclockwise. When your friend observes this situation from a distance, he or she sees SHM and graphs it as shown in Fig. 3-12. Which of the following equations describes this graph?

(a) $q = q_{pk+} \sin (\omega t + \pi/4)$.

(b) $q = q_{pk+} \sin (\omega t + \pi/2)$.

(c) $q = q_{pk+} \sin (\omega t + 3\pi/4)$.

(d) $q = q_{pk+} \sin (\omega t + \pi)$.

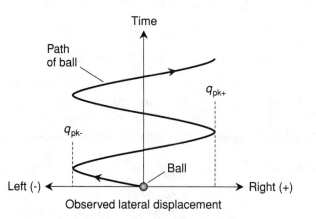

Figure 3-12   Fig. 3-12. Illustration for Quiz Question 1

2. Imagine two disks X and Y, cut from a single sheet of metal having uniform density and uniform thickness. Suppose disk X has a radius of 200 mm while disk Y has a radius of 400 mm. Let I_X be the rotational inertia of disk X relative to its center. Let I_Y be the rotational inertia of disk Y relative to its center. How do these two values compare?

   (a) $I_X = I_Y$.

   (b) $2\,I_X = I_Y$.

   (c) $4\,I_X = I_Y$.

   (d) $16\,I_X = I_Y$.

3. Suppose a point mass Q revolves around a central point in a perfect circle at an angular speed of 4.80 rad/s. The mass of Q is 2.00 kg. What is the tangential momentum vector magnitude?

   (a) 2.40 kg · m/s.

   (b) 9.60 kg · m/s.

   (c) 46.1 kg · m/s.

   (d) We need more information in order to answer this.

4. Imagine an ideal spring-loaded system such as that shown in Fig. 3-11. The cart alone, without its payload, has a mass of 2.00 kg. The payload has a mass of 6.00 kg. The cart and its payload are set in motion, and oscillation occurs at a frequency of 4.00 Hz. Then the payload is suddenly removed, and the empty cart continues to oscillate. What is the new frequency of oscillation?

   (a) 16.0 Hz.

   (b) 8.00 Hz.

   (c) 2.00 Hz.

   (d) 1.00 Hz.

5. In an ideal spring-loaded system such as that shown in Fig. 3-11, a graph of the spring's potential energy as a function of the cart's linear displacement has the general shape of a

   (a) straight line.

   (b) catenary.

   (c) hyperbola.

   (d) parabola.

6. Imagine that you whirl a stone in uniform circular motion on the end of a length of kite line attached to a spool. Then you let out some line with the spool until the line is twice as long, but you make sure that the stone has the same tangential speed as before. What happens to the angular speed of the stone?

(a) It doubles.

(b) It does not change.

(c) It becomes half as great.

(d) We cannot answer this without knowing the original length of the line.

7. Imagine a force **G** of magnitude 12 N, applied at a point Q on the edge of a rotatable disk whose radius is 45 cm. Suppose that **G** lies in the plane of the disk, and dir **G** subtends an angle of $\pi/4$ with respect to the direction of the radius vector **r** extending from the center of the disk to point Q. What is the magnitude r_L of the lever arm?

(a) 32 cm.

(b) 45 cm.

(c) 64 cm.

(d) It is impossible to answer this question without more information.

8. In the scenario of Question 7, what is τ ?

(a) 19 N · m.

(b) 5.4 N · m.

(c) 3.8 N · m.

(d) 7.7 N · m.

9. Suppose a disk has a rotational inertia of 4.50 kg · m². It starts out at an angular speed of 3.00 rad/s, rotating counterclockwise from the point of view of an observer. A constant torque of 0.270 N · m is applied perpendicular to the plane of rotation, and pointing generally away from the observer. How long must this torque be applied in order to bring the disk to a stop?

(a) 50.0 s.

(b) 2.47 s.

(c) 0.180 s.

(d) Forget it! The disk will never stop. It will keep rotating counterclockwise, faster and faster for as long as the torque is applied.

10. In the olden days, vinyl disks were played on turntables that spun at 33⅓ rpm. What was the angular speed of these turntables in radians per second? Consider the value of π to be 3.14.

    (a) 1.74 rad/s.

    (b) 3.49 rad/s.

    (c) 5.65 rad/s.

    (d) 11.3 rad/s.

CHAPTER 4

Thermodynamics

Thermodynamics is a branch of physics that deals with the interaction between *thermal energy* and *mechanical energy*. In this chapter, we'll review the basics of temperature and thermal-energy transfer. Then we'll explore how these phenomena relate to material dimensions, changes of state, and pressure.

Temperature

Temperature, symbolized in equations by T, is a direct or indirect measurement of the amount of thermal energy E_t that exists in matter. Thermal energy is a form of kinetic energy, manifest in the motion of atoms or molecules.

THERMODYNAMIC TEMPERATURE

Thermodynamic temperature is determined according to the average kinetic energy that the atoms or molecules in a substance possess. When the thermodynamic temperature of a sample of matter is low, the atoms or molecules move slowly. As the thermodynamic temperature rises, the atoms or molecules move faster. Because the kinetic energy in a moving particle of constant mass is proportional to the square of its speed, the average kinetic energy in the atoms or molecules of a sample of matter increases as the thermodynamic temperature rises. This phenomenon can be measured by a *thermometer* placed in direct contact with, or immersed in, the sample.

HEAT

Heat is the transfer or flow of thermal energy from one place or material sample to another. When thermal energy is allowed to flow freely from one substance into another, the temperatures of the two media "try to equalize." Ultimately, if the energy transfer process is allowed to continue for a long enough time, the temperatures of the two media become the same unless one of the substances is removed or escapes (for example, steam boiling off of a kettle of water). The tendency of heat to "strive for" uniform temperature throughout a given medium, or to equalize the temperatures of multiple media in contact with each other, is called *entropy*.

DEGREES CELSIUS

The *Celsius temperature scale*, also called the *centigrade temperature scale*, is based on the melting and boiling points of water at normal sea-level atmospheric pressure. Celsius readings are followed by the symbol °C, meaning *degree(s) Celsius*. A change of plus-or-minus one degree Celsius (±1°C) represents 1/100 of the difference between the melting temperature of pure water ice at sea level (0°C) and its boiling temperature at sea level (+100°C).

KELVINS

Temperatures can plunge far below 0°C, and can rise far above +100°C. There is an absolute limit to how low the temperature in degrees Celsius can become, but there is no limit on the upper end of the scale. We can never chill anything down to a temperature lower than approximately −273.15°C. This temperature, which represents the complete absence of thermal energy, is known as *absolute zero*. An object at a temperature of absolute zero can't transfer thermal energy to anything else, because it has none.

Absolute zero is the defining point for the *Kelvin temperature scale*. A temperature of approximately −273.15°C is equal to 0 K (zero kelvins, written without the degree symbol). The kelvin increment is the same as the Celsius increment, so 0°C = 273.15 K, and +100°C = 373.15 K. There is no need to put a plus sign in front of a kelvin figure, because kelvins are never negative.

No matter what the actual temperature of an object, the difference between kelvins and degrees Celsius is always 273.15, with kelvins representing the higher figure.

DEGREES RANKINE

An uncommon scale, called the *Rankine temperature scale*, also assigns the value zero to the coldest possible temperature. The Rankine increment is exactly 5/9 as large as the kelvin increment. Readings in degrees Rankine are followed by the symbol °R. If you want to convert any reading in the Rankine scale to its equivalent in kelvins, multiply by 5/9. Conversely, to convert any figure in kelvins to its equivalent in degrees Rankine, multiply by 9/5 or 1.8. As with kelvins, there is no need to put a plus sign in front of a Rankine figure, because Rankine temperatures are never negative.

DEGREES FAHRENHEIT

In much of the English-speaking world, and especially in the United States, the *Fahrenheit temperature scale* is used by lay people. A Fahrenheit increment is the same size as a Rankine increment, but the scale is situated differently. Readings in degrees Fahrenheit are followed by the symbol °F. The melting temperature of pure water ice at sea level is +32°F, and the boiling point of pure liquid water at sea level is +212°F. A temperature of +32° F corresponds to 0°C, and +212°F corresponds to +100°C. Absolute zero is represented by a reading of approximately −459.67°F.

TEMPERATURE CONVERSIONS

Let C represent the temperature in degrees Celsius, K represent the temperature in kelvins, R represent the temperature in degrees Rankine, and F represent the temperature in degrees Fahrenheit. If you want to convert from one scale to another, you'll find the formula you need among the following:

$$C = (5/9)(F - 32)$$
$$C = K - 273.15$$
$$C = (5/9)(R - 491.67)$$

$$K = C + 273.15$$
$$K = (5/9)\,R$$
$$K = (5/9)\,(F + 459.67)$$

$$R = (9/5)\,(C + 273.15)$$
$$R = (9/5)\,K$$
$$R = F + 459.67$$

$$F = (9/5)\,C + 32$$
$$F = (9/5)\,K - 459.67$$
$$F = R - 459.67$$

The numerical constants 9/5, 5/9, and 32 are mathematically exact for calculation purposes. The other numerical constants are approximated to five significant figures. Don't feel compelled to burn all these formulas into your memory! It's a good idea to memorize how to convert Celsius to Fahrenheit and vice-versa, nevertheless. Figure 4-1 is a nomograph for quick, approximate conversions between these two scales over a range of temperatures commonly encountered, with a few "landmark" points labeled.

PROBLEM 4-1
What are the Celsius, Rankine, and Fahrenheit equivalents of 300 K? Round off the answers to the nearest degree in each scale.

SOLUTION 4-1
These are found using the formulas from above:

$$
\begin{aligned}
C &= K - 273.15 \\
&= 300 - 273.15 \\
&= +27°\text{C}
\end{aligned}
$$

$$
\begin{aligned}
R &= (9/5)\,K \\
&= (9/5) \times 300 \\
&= 540°\text{R}
\end{aligned}
$$

$$
\begin{aligned}
F &= (9/5)\,K - 459.67 \\
&= (9/5) \times 300 - 459.67 \\
&= 540 - 459.67 \\
&= +80°\text{F}
\end{aligned}
$$

PROBLEM 4-2
What are the Celsius, Rankine, and Fahrenheit equivalents of (3.00×10^6) K, a temperature similar to that believed to exist inside of the sun? Express the answers to three significant figures. What happens to the ratio of the Celsius figure to the

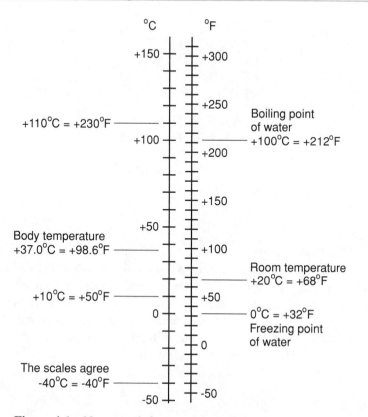

°C   °F

+150 ┼ ┬ +300

           ┼ +250

+110°C = +230°F ────────   Boiling point
                                of water
+100 ┼   ─────── +100°C = +212°F
           ┼ +200

           ┼ +150

+50 ┼
Body temperature
+37.0°C = +98.6°F ──────── ┼ +100
                                Room temperature
                           ─────── +20°C = +68°F
+10°C = +50°F ──────── ┼ +50
           0 ┼   ─────── 0°C = +32°F
                                Freezing point
                           ┼ 0   of water

The scales agree
-40°C = -40°F ────────
         -50 ┼ ┼ -50

Figure 4-1   Nomograph for approximate conversion between
temperature readings in degrees Celsius and degrees Fahrenheit.

Kelvin figure as the temperature increases without limit? What happens to the ratio
of the Celsius figure to the Rankine figure as the temperature increases without
limit? What happens to the ratio of the Celsius figure to the Fahrenheit figure as
the temperature increases without limit?

SOLUTION 4-2

Again, we can use the formulas from above to determine these figures. Let's
carry the intermediate calculations out as far as we can go, and round off only at
the end. Then:

$$C = K - 273.15$$
$$= 3.00 \times 10^6 - 273.15$$
$$= 3,000,000 - 273.15$$
$$= 2,999,726.85°C$$
$$= (2.99972685 \times 10^6)°C$$
$$= (3.00 \times 10^6)°C$$

$$R = (9/5)\ K$$
$$= (9/5) \times 3.00 \times 10^6$$
$$= (9/5) \times 3{,}000{,}000$$
$$= 5{,}400{,}000°R$$
$$= (5.40 \times 10^6)°R$$

$$F = (9/5)\ K - 459.67$$
$$= (9/5) \times 3.00 \times 10^6 - 459.67$$
$$= (9/5) \times 3{,}000{,}000 - 459.67$$
$$= 5{,}400{,}000 - 459.67$$
$$= 5{,}399{,}540.33°F$$
$$= (5.39954033 \times 10^6)°F$$
$$= (5.40 \times 10^6)°F$$

As the temperature increases without limit, the ratio of the Celsius figure to the Kelvin figure approaches 1:1, the ratio of the Celsius figure to the Rankine figure approaches 5:9, and the ratio of the Celsius figure to the Fahrenheit figure also approaches 5:9. At extremely high temperatures, figures in the Celsius and Kelvin scales are practically the same because the difference (273.15) is only a tiny fraction of the numeric value. The same is true of figures in the Fahrenheit and Rankine scales, although scientists rarely use either of them.

Thermal Energy Transfer

When thermal energy is transferred from one object or medium to another, the temperatures and/or states (solid, liquid, or gaseous) of both environments change. The extent and nature of the changes depend on three variables:

- The amount of thermal energy exchanged
- The initial temperatures of the objects or media
- The initial states of the objects or media

HEAT MODES

Thermal energy can be transferred in one or more of three *heat modes* known as *radiation*, *conduction*, and *convection*. The following examples illustrate these phenomena.

When an *infrared* (IR) "heat" lamp shines on your sore shoulder, thermal energy is transferred to your skin surface by radiation from the hot filament. If you place a kettle of water on a hot stove, thermal energy is transferred by conduction from the burner to the water. When a fan-type electric heater warms a room, air passes through the heating elements and is blown into the room where the heated air rises and mixes by convection with the rest of the air in the room.

In an old-fashioned wood stove, a controlled fire heats an iron box, which in turn emits thermal energy as IR radiation that warms the walls, floor, ceiling, and furniture in the room. Thermal energy is transferred to the air by conduction from the hot stove and from the warm walls, floor, ceiling, and furniture. The warmed air rises, causing continuous convection that helps to equalize the temperature throughout the room. A wood stove therefore takes advantage of all three heat modes (Fig. 4-2).

THE JOULE

Physicists measure and express energy, regardless of its form, in units called *joules*. One joule (1 J) represents one *watt* (1 W) of *power* expended, radiated, or

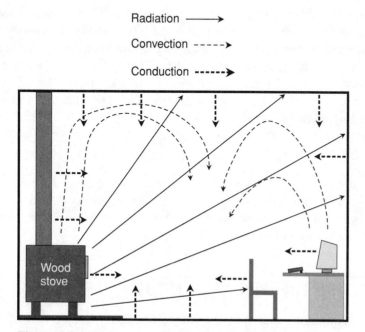

Figure 4-2   A wood stove heats a room by all three modes of thermal energy transfer: radiation, convection, and conduction.

dissipated for one second (1 s) of time. A joule is therefore equivalent to a *watt-second*, and a watt is the equivalent of a *joule per second*. Mathematically:

$$1 \text{ J} = 1 \text{ W} \cdot \text{s}$$
$$1 \text{ W} = 1 \text{ J/s} = 1 \text{ J} \cdot \text{s}^{-1}$$

The joule is the equivalent of a *kilogram meter squared per second squared* in base Standard International (SI) units:

$$1 \text{ J} = 1 \text{ kg} \cdot \text{m}^2/\text{s}^2$$

In order to get an idea of how much heat a joule represents, imagine an incandescent lamp typical of utility-powered "night lights" that is rated at 5.00 W. If the lamp is 20.0% efficient (a typical figure for small incandescent bulbs), then it radiates 1.00 W as visible light and 4.00 W as *thermal power*, which is the equivalent of 4.00 J of thermal energy per second, 240 J of thermal energy per minute, or 14.4 kJ of thermal energy per hour. That's nowhere near enough thermal energy to heat up a cold room, but more than enough to burn your fingers if you grab the bulb and hold on!

THE CALORIE

The *calorie* is another unit of heat, less often used. One calorie (1 cal) is the amount of heat that raises the temperature of 1 g of pure liquid water by 1°C. It is also the amount of thermal energy lost by 1 g of pure liquid water if its temperature falls by 1°C. The *kilocalorie* (kcal), also called a *diet calorie*, is the amount of energy transfer involved when the temperature of 1 kg, or 1000 g, of pure liquid water rises or falls by 1°C. The calorie and the joule are related as follows, accurate to four significant figures:

$$1.000 \text{ cal} = 4.184 \text{ J}$$
$$1.000 \text{ J} = 0.2390 \text{ cal}$$

This definition of the calorie holds strictly true only as long as the water remains in the liquid state during the entire thermal-energy-transfer process. At standard atmospheric pressure on the earth's surface, it holds for temperatures above 0°C and below +100°C.

THE BRITISH THERMAL UNIT (BTU)

In home heating applications in the United States, an archaic unit of energy is used: the *British thermal unit* (Btu). You'll hear this unit mentioned in advertisements for furnaces and air conditioners. One British thermal unit (1 Btu) is the amount of energy transfer that raises the temperature of one pound (1 lb) of pure liquid water by 1°F. It is also the amount of energy lost by 1 lb of pure liquid water if its temperature falls by 1°F. This definition, like that of the calorie, holds strictly true only between the freezing and boiling points.

When people talk about "Btus" literally in regards to the heating or cooling capacity of a furnace or air conditioner, they're using the term improperly. They really mean to quote the thermal power in *British thermal units per hour* (Btu/h), not the total amount of heat in British thermal units. The British thermal unit is related to the joule as follows, accurate to four significant figures:

$$1.000 \text{ Btu} = 1055 \text{ J}$$
$$1.000 \text{ J} = 9.479 \times 10^{-4} \text{ Btu}$$

This holds strictly true only as long as the water remains in the liquid state during the entire thermal-energy-transfer process. At standard atmospheric pressure on the earth's surface, therefore, it is valid for temperatures above +32°F and below +212°F.

HEAT CAPACITY

Heat capacity, symbolized C, is the ratio of the thermal energy ΔE_t absorbed or given up by a sample to the extent ΔT by which its temperature rises or falls. This is true as long as the sample does not change state:

$$C = \Delta E_t / \Delta T$$

where C is in *joules per kelvin* (J/K), ΔE_t is in joules, and ΔT is in kelvins. Alternatively, the formula works for C in *calories per kelvin* (cal/K), ΔE_t in calories, and ΔT in kelvins. (A change of state—solid-to-liquid, liquid-to-solid, liquid-to-gas, or gas-to-liquid—makes a big difference, as we will shortly see.) Sometimes C is given in *joules per degree Celsius* (J/°C) or *calories per degree Celsius* (cal/°C).

Specific heat capacity, also known as *specific heat*, is expressed in *joules per kilogram per kelvin* (J/kg/K) or *calories per gram per degree Celsius* (cal/g/°C) and is symbolized in equations by the lowercase italic letter c. Don't confuse this with the speed of light in free space, also symbolized by c! The first definition represents the number of joules it takes to raise or lower the temperature of 1 kg of a substance by 1 K when there is no change of state. The second definition represents the number of calories it takes to raise or lower the temperature of 1 g of a substance by 1°C when there is no change of state. Mathematically:

$$c = \Delta E_t / (m \, \Delta T)$$

where c is in joules per kilogram per kelvin, ΔE_t is in joules, m is in kilograms, and ΔT is in kelvins. Alternatively, the formula works for c in calories per gram per degree Celsius, ΔE_t in calories, m in grams, and ΔT in degrees Celsius.

Molar heat capacity is expressed in *joules per mole per kelvin* (J/mol/K) or *calories per mole per degree Celsius* (cal/mol/°C) and is symbolized in equations as C_m. The first definition represents the number of joules it takes to raise or lower the temperature of one *mole* (1 mol) of a substance by 1 K when there is no change of state. In this context, one mole of a substance represents about 6.02×10^{23} atoms of that substance. This large number comes up frequently in chemistry, is symbolized N_A in equations, and is known as *Avogadro's number* or *Avogadro's constant*. According to NIST, its value to six significant figures is:

$$N_A = 6.02214 \times 10^{23} \text{ /mol}$$

where /mol, also written mol^{-1}, is read "per mole." The second definition represents the number of calories it takes to raise or lower the temperature of 1 mol of a substance by 1°C when there is no change of state. Mathematically:

$$C_m = \Delta E_t / (N \, \Delta T)$$

where ΔE_t is in joules or calories, N is in moles, and ΔT is in kelvins or degrees Celsius.

HEAT OF FUSION

It takes a certain amount of heat to change a sample of solid matter to its liquid state or vice-versa, assuming the matter is of the sort that can exist in either of these two states. In dry air at sea level, 3.35×10^5 J are required to convert 1.00 kg of pure water ice at 0°C to 1.00 kg of pure liquid water at 0°C. In the reverse scenario, if 1.00 kg of pure liquid water at 0°C freezes completely solid and becomes 1.00 kg of ice at 0°C, it gives up 3.35×10^5 J. If you want to work with calories and grams instead of joules and kilograms, it takes 80.0 cal to convert 1.00 g of pure water ice at 0°C to 1.00 g of pure liquid water at 0°C, or vice-versa.

This parameter is called the *heat of fusion*, and its value varies greatly depending on the substance involved. Heat of fusion can be expressed in joules per kilogram (J/kg) or calories per gram (cal/g). When the substance is something other than water, the freezing/melting point of that substance must be substituted for 0°C.

If the heat of fusion is symbolized h_f, the thermal energy added or given up by a sample of matter is ΔE_t, and the mass of the sample is m, then the following formula holds:

$$h_f = \Delta E_t / m$$

where h_f is in joules per kilogram, ΔE_t is in joules, and m is in kilograms. The formula also holds for h_f in calories per gram, ΔE_t in calories, and m in grams.

HEAT OF VAPORIZATION

A certain amount of heat is necessary in order for a sample of liquid to boil off and become gaseous, assuming the matter is of the sort that can exist in either of these two states. In dry air at sea level, it takes 2.26×10^6 J to convert 1.00 kg of pure liquid water at +100°C to 1.00 kg of pure water vapor at +100°C. If 1.00 kg of pure water vapor at +100°C condenses completely and becomes liquid at +100°C, it gives up 2.26×10^6 J. Alternatively, it takes 540 cal to convert 1.00 g of pure liquid water at +100°C to 1.00 g of pure vapor at +100°C, or vice-versa.

This parameter, like heat of fusion, varies for different substances, and is called the *heat of vaporization*. It, like heat of fusion, is expressed in joules per kilogram or calories per gram. When the substance is something other than water at sea level, then the boiling or condensation point of that substance must be substituted for +100°C.

If the heat of vaporization is symbolized h_v, the thermal energy added or given up by a sample of matter is ΔE_t, and the mass of the sample is m, then the following formula holds:

$$h_v = \Delta E_t / m$$

where h_v is in joules per kilogram, ΔE_t is in joules, and m is in kilograms. The formula also holds for h_f in calories per gram, ΔE_t in calories, and m in grams.

PROBLEM 4-3
Suppose a certain liquid has a specific heat of 1.50 cal/g/°C. If a 65.5-g sample of this substance gives up 765 cal of heat, and assuming it remains liquid, how much will it cool off? Express the answer in degrees Celsius to three significant figures.

SOLUTION 4-3
Use the formula for specific heat c in terms of energy change ΔE_t, mass m, and temperature change ΔT, as follows:

$$c = \Delta E_t / (m \, \Delta T)$$

where c is in calories per gram per degree Celsius, ΔE_t is in calories, m is in grams, and ΔT is in degrees Celsius. We can rearrange this formula to obtain ΔT in terms of the other variables:

$$\Delta T = \Delta E_t / (cm)$$

We are given the values $\Delta E_t = -765$, $c = 1.50$, and $m = 65.5$. The change in energy is negative because the sample gives up energy. Therefore:

$$\Delta T = -765 \,/\, (1.50 \times 65.5)$$
$$= -765 \,/\, 98.25$$
$$= -7.79°C$$

The negative figure for the result indicates that the temperature falls.

Thermal Behavior of Solids

Most solids expand as the temperature rises and contract as the temperature falls. A few substances contract as they become warmer and expand as they become cooler. The extent of such expansion and contraction is rarely more than a tiny fraction of the actual size of a solid object. We can quantify *thermal expansion* and *thermal contraction* according to linear dimension, surface area, or volume.

LINEAR EXPANSION AND CONTRACTION

The extent to which the height, width, or depth of a solid (its *linear dimension*) changes per degree change in temperature when all other factors are held constant is called the *thermal coefficient of linear expansion*. For most materials, the coefficient of linear expansion maintains the same value within a reasonable range of temperatures. If the temperature rises by 2°C, for instance, the linear dimension of a length of steel wire increases twice as much as it would if the temperature increase were only 1°C, provided the wire doesn't soften or melt, and as long as no other factors such as mechanical stress are present.

Coefficient of linear expansion is usually defined in meters per meter per kelvin, or in meters per meter per degree Celsius. Meters cancel out in these expressions of units, so the technical quantities are *per kelvin* (/K) or *per degree Celsius* (/°C). These units are equivalent because the kelvin and Celsius increments are the same.

If Δs is the difference in linear dimension produced by a temperature change of ΔT for a solid object whose initial linear dimension is s, then the thermal coefficient of linear expansion, symbolized by the lowercase Greek letter alpha with the subscript s (α_s), is given by this equation:

$$\alpha_s = \Delta s \,/\, (s\, \Delta T)$$

where α_s is in per kelvin or per degree Celsius, s and Δs are in meters, and ΔT is in kelvins or degrees Celsius. Positive values of α_s indicate that the linear dimension increases as the temperature rises. Negative values of α_s indicate that the linear dimension decreases as the temperature rises. When the linear dimension increases, Δs is considered positive; when it decreases, Δs is considered negative. Rising temperatures are expressed as positive values of ΔT, while falling temperatures are expressed as negative values of ΔT.

SURFACE-AREA EXPANSION AND CONTRACTION

The *thermal coefficient of surface-area expansion* is defined in meters squared per meter squared per kelvin, or in meters squared per meter squared per degree Celsius. Meters squared cancel out in these expressions of units, so the technical quantities are per kelvin (/K) or per degree Celsius (/°C), the same units as for the coefficient of linear expansion.

If ΔA is the difference in surface area produced by a temperature change of ΔT for a solid object whose initial surface area is A, then the thermal coefficient of surface-area expansion, symbolized by the lowercase Greek letter alpha with the subscript a (α_a), is given by this equation:

$$\alpha_a = \Delta A \,/\, (A \,\Delta T)$$

where α_a is in per kelvin or per degree Celsius, A and ΔA are in meters squared, and ΔT is in kelvins or degrees Celsius. Positive values of α_a indicate that the surface area increases as the temperature rises. Negative values of α_a indicate that the surface area decreases as the temperature rises.

If the actual extent of expansion or contraction is small compared with the initial dimensions, the value of α_a is almost exactly twice the value of α_s for any particular solid substance. (Actually it is a tiny bit more than twice, but the difference is negligible in most real-world scenarios.) When the surface area increases, ΔA is positive; when the surface area decreases, ΔA is negative.

VOLUME EXPANSION AND CONTRACTION

The *thermal coefficient of volume expansion* is defined in meters cubed per meter cubed per kelvin, or in meters cubed per meter cubed per degree Celsius. Meters cubed cancel out in these expressions, so the technical quantities are, once again, per kelvin (/K) or per degree Celsius (/°C).

In general, if ΔV is the difference in volume produced by a temperature change of ΔT for a solid object whose initial volume is V, then the thermal coefficient of volume expansion, symbolized by the lowercase Greek letter alpha with the subscript v (α_v), is given by this equation:

$$\alpha_v = \Delta V / (V \, \Delta T)$$

where α_v is in per kelvin or per degree Celsius, V and ΔV are in meters cubed, and ΔT is in kelvins or degrees Celsius. Positive values of α_v indicate that the volume increases as the temperature rises. Negative values of α_v indicate that the volume decreases as the temperature rises. In some texts, coefficient of volume expansion is symbolized by the lowercase Greek letter beta (β).

If the actual extent of expansion or contraction is small compared with the initial dimensions, the value of α_v is almost exactly three times the value of α_s for any particular solid substance. (Actually it is a tiny bit more than three times, but the difference is negligible in most real-world scenarios.) When the volume increases, ΔV is positive; when the volume decreases, ΔV is negative.

PROBLEM 4-4

Consider a block of metal whose thermal coefficient of linear expansion is $+1.2500 \times 10^{-5}$ /K. Suppose the temperature rises from 10.000°C to 30.000°C, the block remains solid, and it does not soften. At 10.000°C, the block is 30.000 cm tall by 40.000 cm wide by 20.000 cm deep. To four significant figures, find:

1. The extent to which the block's height changes with the increase in temperature
2. The extent to which the block's width changes with the increase in temperature
3. The extent to which the block's depth changes with the increase in temperature
4. The extent to which the block's surface area changes with the increase in temperature
5. The extent to which the block's volume changes with the increase in temperature

SOLUTION 4-4

1. Let's call the initial height s_h. We are told that $s_h = 30.000$ cm. Converting to meters, we get $s_h = 0.30000$ m. We are also told that $\alpha_s = +1.2500 \times 10^{-5}$. The temperature rises by exactly 20.000°C, so $\Delta T = +20.000$. Let's call the change in height Δs_h. Rearranging the formula for the coefficient of linear expansion with s_h in place of s and Δs_h in place of Δs, we obtain:

$$\alpha_s = \Delta s_h / (s_h \, \Delta T)$$
$$\Delta s_h = \alpha_s \, s_h \, \Delta T$$
$$= +1.2500 \times 10^{-5} \times 0.3000 \times (+20.000)$$
$$= +7.500 \times 10^{-5} \text{ m}$$

The plus sign indicates that the block grows taller.

2. Let's call the initial width s_w. We are told that s_w = 40.000 cm. Converting to meters, we get s_w = 0.40000 m. We know that α_s = +1.2500 × 10⁻⁵ and ΔT = +20.000. Let's call the change in width Δs_w. Rearranging the formula derived above with s_w in place of s_h and Δs_w in place of Δs_h, we obtain:

$$\Delta s_w = \alpha_s \, s_w \, \Delta T$$
$$= +1.2500 \times 10^{-5} \times 0.4000 \times (+20.000)$$
$$= +1.000 \times 10^{-4} \text{ m}$$

The plus sign indicates that the block grows wider.

3. Let's call the initial depth s_d. We are told that s_d = 20.000 cm. Converting to meters, we get s_d = 0.20000 m. We know that α_s = +1.2500 × 10⁻⁵ and ΔT = +20.000. Let's call the change in depth Δs_d. Rearranging the formula derived above with s_d in place of s_h and Δs_d in place of Δs_h, we obtain:

$$\Delta s_d = \alpha_s \, s_d \, \Delta T$$
$$= +1.2500 \times 10^{-5} \times 0.2000 \times (+20.000)$$
$$= +5.000 \times 10^{-5} \text{ m}$$

The plus sign indicates that the block grows deeper.

4. The coefficient of surface-area expansion α_a is approximately twice the coefficient of linear expansion α_s. In this case:

$$\alpha_a = 2 \, \alpha_s$$
$$= 2.0000 \times (+1.2500 \times 10^{-5})$$
$$= +2.5000 \times 10^{-5}$$

Let's call the initial surface area A. We can determine this on the basis of the known initial height s_h, the known initial width s_w, and the known initial depth s_d, as follows:

$$A = 2 \, s_h \, s_w + 2 \, s_w \, s_d + 2 \, s_h \, s_d$$
$$= 2.0000 \times 0.3000 \times 0.4000$$
$$+ 2.0000 \times 0.4000 \times 0.2000$$
$$+ 2.0000 \times 0.3000 \times 0.2000$$
$$= 0.2400 + 0.1600 + 0.1200$$
$$= 0.5200 \text{ m}^2$$

We know that α_a = +2.5000 × 10⁻⁵. We also know that ΔT = +20.000. Rearranging the formula for the coefficient of surface-area expansion, we can solve for ΔA, the change in surface area:

$$\alpha_a = \Delta A \, / \, (A \, \Delta T)$$
$$\Delta A = \alpha_a \, A \, \Delta T$$
$$= +2.5000 \times 10^{-5} \times 0.5200 \times (+20.000)$$
$$= +2.600 \times 10^{-4} \text{ m}^2$$

The plus sign indicates that the surface area of the block increases.

5. The coefficient of volume expansion α_v is approximately three times the coefficient of linear expansion α_s. In this case:

$$\begin{aligned} \alpha_v &= 3\,\alpha_s \\ &= 3.0000 \times (+1.2500 \times 10^{-5}) \\ &= +3.7500 \times 10^{-5} \end{aligned}$$

Let's call the initial volume V. We can determine this on the basis of the known initial height s_h, the known initial width s_w, and the known initial depth s_d, as follows:

$$\begin{aligned} V &= s_h\, s_w\, s_d \\ &= 0.3000 \times 0.4000 \times 0.2000 \\ &= 0.02400 \text{ m}^3 \end{aligned}$$

We know that $\alpha_v = +3.7500 \times 10^{-5}$ and $\Delta T = +20.000$. Rearranging the formula for the coefficient of volume expansion, we can solve for ΔV, the change in volume:

$$\begin{aligned} \alpha_v &= \Delta V / (V\,\Delta T) \\ \Delta V &= \alpha_v\, V\, \Delta T \\ &= +3.7500 \times 10^{-5} \times 0.02400 \times (+20.000) \\ &= +1.800 \times 10^{-5} \text{ m}^3 \end{aligned}$$

The plus sign indicates that the volume of the block increases.

Thermal Behavior of Gases

The gaseous phase of matter is similar to the liquid phase, insofar as a gas will conform to the boundaries of a container or enclosure. But a gas is much less affected by gravity than a liquid. If you fill up a bottle with a gas, there is no discernible "surface" to the gas. Another difference between liquids and gases is the fact that gases are compressible, while liquids generally aren't.

GAS DENSITY

Gas density can be defined in three ways. *Mass density* (D_m) is expressed in kilograms per meter cubed (kg/m³) for a sample of gas. *Weight density* (D_w) is expressed in newtons per meter cubed (N/m³), and is equal to the mass density multiplied by the acceleration in meters per second squared (m/s²) to which the

sample is subjected. *Particle density* (D_p) is the number of moles of atoms per meter cubed (mol/m^3) in a parcel or sample of gas, where 1 mol is Avogadro's constant N_A, approximately 6.02×10^{23}.

GAS PRESSURE

Gas pressure, symbolized by the uppercase italic letter P in equations, is expressed in terms of the force magnitude per unit area that a gas exerts on the walls of a container in newtons per meter squared (N/m^2). The newton per meter squared is a standard unit of pressure also known as the *Pascal* (Pa).

Imagine a container whose volume (in meters cubed) is equal to V. Suppose there are N moles of atoms of a particular gas inside this container, which is surrounded by a perfect vacuum. We can say certain things about the pressure P (in pascals) that the gas exerts outward on the walls of the container. First, P is proportional to N, provided that V is held constant. Second, if V increases while N remains constant, P decreases. These things are intuitively apparent. But there's more to the story, as we will soon see.

GAS TEMPERATURE

Gas temperature, symbolized by the uppercase italic letter T in equations, is commonly expressed in kelvins, because this scale has its zero point at absolute zero. However, gas temperature is occasionally specified in degrees Celsius or degrees Fahrenheit. When a parcel of gas is compressed, its temperature rises; when it is decompressed, its temperature falls. If all other factors are held constant, heating up a parcel of gas increases the pressure, and cooling it off reduces the pressure.

STANDARD TEMPERATURE AND PRESSURE (STP)

To set a reference for temperature and pressure for gases against which measurements can be made and experiments conducted, scientists have defined *standard temperature and pressure* (STP). This represents the typical state of affairs at sea level on the earth's surface. The standard temperature is 0°C (32°F). Standard pressure is the air pressure that will support a column of liquid elemental mercury 0.760 m tall; it turns out to be approximately 1.01×10^5 Pa on our planet, and is often called *one atmosphere* (1 atm). In American lay people's terms, standard pressure is about 14.7 pounds per inch squared (lb/in^2).

IDEAL GAS LAW

An *ideal gas*, also called a *perfect gas*, is a theoretical system in which the fundamental particles (atoms or molecules) have vanishingly small volume, no forces exist among them, and any collision that happens to take place between two particles is perfectly elastic.

Let P be the pressure of an ideal gas having absolute temperature T, confined to an enclosure having volume V. Let N represent the number of atoms or molecules of gas in the enclosure. Then the following relationships hold among these variables:

$$P = 8.31 \, NT / V$$
$$T = PV / (8.31 \, N)$$
$$V = 8.31 \, NT / P$$
$$N = PV / (8.31 \, T)$$

where P is in pascals, T is in kelvins, V is in meters cubed, N is in moles, and 8.31 is the *universal gas constant*, sometimes symbolized by the uppercase italic letter R and defined in *joules per mole per kelvin* (J/mol/K). The value of this constant, also known as the *molar gas constant*, is given by NIST to six significant figures as:

$$R = 8.31447 \text{ J/mol/K}$$

The equations above are statements of a fundamental rule of gas behavior known as the *ideal gas law*.

THREE COROLLARIES

The ideal gas law implies three significant corollaries, all of which can be derived directly from the equations above.

Corollary 1. When the pressure exerted by a *confined dry gas* (that is, a sample of matter that contains a constant number of atoms or molecules and that remains entirely gaseous at all times) is constant, the volume is directly proportional to the absolute temperature:

$$V = k_1 T$$

where V is the volume, T is the absolute temperature in kelvins, and k_1 is a constant that depends on the volume units we specify, and on the overall number of gas atoms or molecules. This is called *Charles's law*. If the pressure P is in pascals, V is in meters cubed, and N is in moles, then:

$$k_1 = 8.31 \, N / P$$
$$= RN / P$$

Corollary 2. When the temperature of a confined dry gas is constant, the volume is inversely proportional to the pressure:

$$V = k_2 / P$$

where V is the volume, P is the pressure, and k_2 is a constant that depends on the volume and pressure units we specify, and on the overall number of gas atoms or molecules. This is known as *Boyle's law.* If V is in meters cubed, P is in pascals, N is in moles, and T is the constant temperature in kelvins, then:

$$k_2 = 8.31\ NT$$
$$= RNT$$

Corollary 3. When the volume of a confined dry gas is constant, the absolute temperature is directly proportional to the pressure:

$$T = k_3 P$$

where T is the absolute temperature in kelvins, P is the pressure, and k_3 is a constant that depends on the pressure units we specify, and the overall number of gas atoms or molecules. If P is in pascals, V is in meters cubed, and N is in moles, then:

$$k_3 = V / (8.31\ N)$$
$$= V / (RN)$$

BOLTZMANN'S CONSTANT

In a sample of gas, the individual atoms or molecules move at different speeds, and the instantaneous speed of any single atom or molecule can vary with time. It's impossible to scrutinize the behavior of one particle and then characterize the behavior of the whole sample on that particle alone. However, the average kinetic energy of all the particles in a sample of gas is predictable and follows rules we can quantify.

The mean (average) kinetic energy E_{km} of the atoms or molecules in a dry gas is directly proportional to the absolute thermodynamic temperature T, as follows:

$$E_{km} = k_4 T$$

where k_4 is a constant that depends on the units of kinetic energy and temperature we specify. For a pure, dry *monatomic gas* such as helium, neon, or argon (in which each molecule contains one atom), if E_{km} is given in joules and T is in kelvins, the value of k_4 is approximately 2.07×10^{-23}. This happens to be 3/2 times the ratio of the universal gas constant R to Avogadro's constant N_A. Therefore:

$$E_{km} = (3/2) (R / N_A) T$$

The quantity R / N_A is commonly known as *Boltzmann's constant*. It is expressed in joules per kelvin (J/K) and is symbolized k_b in equations. (Some texts simply use k without any subscript.) Therefore:

$$E_{km} = (3/2) \, k_b \, T$$

Boltzmann's constant is equal to approximately 1.38×10^{-23} J/K. If you want more accuracy, you can use the NIST value as given to six significant figures:

$$k_b = 1.38065 \times 10^{-23} \text{ J/K}$$

For gases in which each molecule contains more than one atom, the situation is more complicated because the molecules contain kinetic energy inherent in their rotations as well as in their movements relative to each other. In the case of a pure, dry *diatomic gas* such as hydrogen or oxygen (in which each molecule contains two atoms), the above equation becomes approximately:

$$E_{km} = (5/2) \, k_b \, T$$

Henceforth, we will deal only with ideal monatomic gases, so we need not be concerned with the energy contributed by the rotation of multi-atom molecules.

ROOT-MEAN-SQUARE SPEED OF GAS PARTICLES

Just as it's impractical to talk about the instantaneous kinetic energy of any single atom or molecule of gas in a sample, there's little point in trying to figure out the speed of any individual particle at a specific point in time. But we can quantify the *mean* (arithmetic average) speed of all the particles in an enclosed sample of gas.

Recall the formula for the kinetic energy E_k of an object in terms of its mass m and its linear speed v, as follows:

$$E_k = m \, v^2 / 2$$

where E_k is in joules, m is in kilograms, and v is in meters per second. Let E_{km} represent the mean kinetic energy of the atoms or molecules in a sample of gas, while v_m represents their mean speed. Then the above formula can be generalized to become:

$$E_{km} = m \, v_m^2 / 2$$

We also know the following equation from the preceding section about Boltzmann's constant:

$$E_{km} = (3/2) \, k_b \, T$$

We can substitute the value on the right-hand side of the first equation for E_{km} in the second equation to get:

$$m \, v_m^2 \, / \, 2 = (3/2) \, k_b \, T$$

This can be manipulated to obtain:

$$v_m^2 = 3 \, k_b \, T \, / \, m$$

If we take the positive square root (or 1/2 power) of both sides of this equation, we get:

$$(v_m^2)^{1/2} = (3 \, k_b \, T \, / \, m)^{1/2}$$

where k_b is Boltzmann's constant, T is in kelvins, m is in kilograms, and $(v_m^2)^{1/2}$ is a special expression of *effective mean speed* known as the *root-mean-square (rms) speed*. Scientists often use rms values when they want to express the effective behavior of a parameter that is variable or unpredictable on a small scale or from moment-to-moment, but that is quantifiable and predictable on a large scale or over a period of time.

Root-mean-square speed is symbolized v_{rms} so the above equation can be rewritten this way:

$$v_{rms} = (3 \, k_b \, T \, / \, m)^{1/2}$$

If all the atoms or molecules in a sample of gas have the same mass and the temperature is uniform throughout the sample, then the rms speed of the particles is directly proportional to the square root of the absolute temperature.

ZEROTH LAW OF THERMODYNAMICS

When two samples of matter are brought into contact so thermal energy can freely flow between them, they will eventually attain a state in which the flow of thermal energy from one to the other is zero. This state, called *thermal equilibrium*, will be reached within a finite time if the samples have finite mass and finite volume. When two samples of matter are in thermal equilibrium, they have the same temperature. This rule, usually called the *zeroth law of thermodynamics*, has two corollaries.

Corollary 1. Consider three objects or samples of matter called X, Y, and Z. If X is in thermal equilibrium with Y, and Y is in thermal equilibrium with Z, then X is in thermal equilibrium with Z.

Corollary 2. Consider a sample of matter having finite mass and finite volume in which thermal energy can freely flow, and in which the initial thermal energy distribution is not uniform. Then thermal energy will flow from the warmer regions to the cooler regions until the temperature is uniform throughout the sample.

PROBLEM 4-5

Imagine a container of pure, dry neon gas. Suppose the temperature is exactly 0°C. Consider the mass of a neon molecule, which is monatomic, to be 1.66×10^{-26} kg. What is the rms speed of the molecules? Express the answer to three significant figures.

SOLUTION 4-5

We must convert the temperature to kelvins. We are told that the temperature is exactly 0°C, so we can convert this to $T = 273.15$ K. Boltzmann's constant is $k_b = 1.38 \times 10^{-23}$ J/K. Using the equation for rms speed in terms of temperature and mass, we calculate:

$$
\begin{aligned}
v_{rms} &= (3 \, k_b \, T \, / \, m)^{1/2} \\
&= [3.00 \times 1.38 \times 10^{-23} \times 273.15 \, / \, (1.66 \times 10^{-26})]^{1/2} \\
&= (6.8123 \times 10^5)^{1/2} \\
&= 825 \text{ m/s}
\end{aligned}
$$

PROBLEM 4-6

What is the rms kinetic energy, E_{krms}, of the molecules in the scenario of Problem 4-5? How does this compare with the mean kinetic energy, E_{km}? Express the answers to two significant figures.

SOLUTION 4-6

To determine the rms kinetic energy, we can use the equation from classical mechanics. In order to avoid cumulative rounding errors, let's use the figure we derived for v_{rms} before we took the square root:

$$
\begin{aligned}
E_{krms} &= m \, v_{rms}^2 \, / \, 2 \\
&= 1.66 \times 10^{-26} \times [(6.8123 \times 10^5)^{1/2}]^2 \, / \, 2 \\
&= 1.66 \times 10^{-26} \times 6.8123 \times 10^5 \, / \, 2 \\
&= 5.7 \times 10^{-21} \text{ J}
\end{aligned}
$$

In order to determine the mean kinetic energy of the molecules, we can use the formula derived earlier in the section on Boltzmann's constant:

$$
\begin{aligned}
E_{km} &= (3/2) \, k_b \, T \\
&= 1.5 \times 1.38 \times 10^{-23} \times 273.15 \\
&= 5.7 \times 10^{-21} \text{ J}
\end{aligned}
$$

In this case, the mean and the rms kinetic-energy values turn out to be the same. This is often, but not always, true in real-world situations. We'll encounter rms values again in Chapter 7 when we study alternating current (AC).

PROBLEM 4-7

What would happen to the rms speed, the rms kinetic energy, and the mean kinetic energy of the molecules in the above situation if the temperature were reduced to absolute zero?

SOLUTION 4-7

In that situation, we would have $T = 0$ (exactly). When we calculate the rms speed of the molecules, we obtain:

$$v_{rms} = (3 \, k_b \, T / m)^{1/2}$$
$$= [3.00 \times 1.38 \times 10^{-23} \times 0 / (1.66 \times 10^{-26})]^{1/2}$$
$$= 0 \text{ m/s (exactly)}$$

Calculating the rms kinetic energy, we get:

$$E_{krms} = m \, v_{rms}^2 / 2$$
$$= 1.66 \times 10^{-26} \times 0^2 / 2$$
$$= 0 \text{ J (exactly)}$$

Calculating the mean kinetic energy, we get:

$$E_{km} = (3/2) \, k_b \, T$$
$$= 1.5 \times 1.38 \times 10^{-23} \times 0$$
$$= 0 \text{ J (exactly)}$$

Significant digits don't matter in these calculations if the absolute temperature is theoretically zero. These results tell us that if there were no thermal energy in a system, the atoms or molecules would not move, and their rms and mean kinetic energy would vanish.

Thermal Processes and Cyclic Heat Engines

A gas can be subjected to thermal processes that involve changes in its volume, pressure, temperature, and thermal energy. In the real world these processes interact, and can be used to design and build systems that convert thermal energy to useful work by going through a cycle. Such a device or machine is called a *cyclic heat engine*.

PRESSURE-VOLUME-TEMPERATURE (PVT) COORDINATES

Three characteristics of any gas can be varied and measured directly: pressure (*P*), volume (*V*), and absolute temperature (*T*). When they are portrayed on the axes in *rectangular three-space* as shown in Fig. 4-3, these parameters form the basis for a *pressure-volume-temperature* (PVT) *system of coordinates*, also called *PVT space*. We can identify three special types of two-dimensional subsets of *PVT* space:

- Any plane perpendicular to the *P* axis consists of points at which the pressure *P* is constant. Such a plane is called an *isobar*.
- Any plane perpendicular to the *V* axis consists of points at which the volume *V* is constant. Such a plane is called an *isochor*.
- Any plane perpendicular to the *T* axis consists of points at which the temperature *T* is constant. Such a plane is called an *isotherm*.

PRESSURE-VOLUME (PV) COORDINATES

The thermal behavior of gases is difficult to illustrate in three dimensions, although animated computer graphics can do a fair job. On paper, lines and curves can be plotted in a simplified system of coordinates called a *PV diagram* that por-

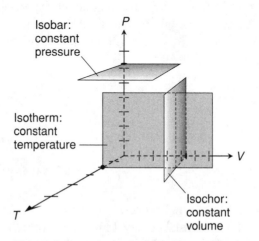

Figure 4-3   A three-dimensional *PVT* coordinate system on which pressure *P*, volume *V*, and temperature *T* can be plotted, showing an isobar, an isochor, and an isotherm.

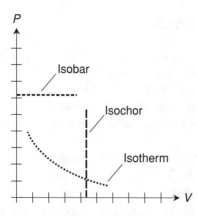

Figure 4-4   A two-dimensional PV coordinate system on which pressure P and volume V can be plotted, showing "slices" of an isobar, an isochor, and an isotherm.

trays relationships between pressure and volume. The result is a two-dimensional "snapshot" of the PVT coordinate system as it would look from a vantage point somewhere far out along the T axis.

Figure 4-4 shows a PV diagram in which two-dimensional "snapshots" of an isobar, an isochor, and an isotherm are plotted. In this two-dimensional diagram, isobars always appear parallel to the V axis as straight horizontal lines. Isochors always appear parallel to the P axis as straight vertical lines. Isotherms can show up as straight lines having any orientation, or as curves.

ISOBARIC PROCESSES

Imagine a process that takes place, or is imposed upon, an inflated balloon in such a way that the pressure inside the balloon stays constant. This is called an *isobaric process* and can be portrayed as any line or curve in PVT space, as long as it lies entirely within a single isobar plane. If graphed in a PV diagram, an isobaric process always appears as a straight, horizontal line because we see the isobar plane edge-on.

The temperature of a balloon can be increased, but if the balloon expands to the right extent, the pressure inside does not change. Air can be let out of the balloon, but if it shrinks to the right extent, again, the pressure stays the same. Anything can be done to the balloon or the gas inside, but as long as the pressure inside the balloon doesn't change, the process is isobaric.

ISOCHORIC PROCESSES

Now envision a process that takes place, or is imposed upon, an inflated balloon so that its volume does not change. Such an event is called an *isochoric process* or *isovolumetric process*. In *PVT* space, it can appear as any line or curve that lies entirely within a single isochor plane. If graphed in a *PV* diagram, the process always appears as a straight, vertical line because we see the isochor plane from an edgewise point of view.

The temperature of a balloon can be increased, but just enough of the gas can be allowed to escape so the balloon stays inflated to the same size. The balloon can be pumped up and cooled at the same time so the gas inside becomes more dense, again keeping it at the same size. Anything can be done to a gas-filled enclosure, but as long as the volume doesn't change, the process is isochoric.

ISOTHERMIC PROCESSES

Consider some process that takes place, or is imposed upon, an inflated balloon in such a way that the temperature inside the balloon stays constant. This is called an *isothermal process* or *isothermic process* and can be portrayed as any line or curve in *PVT* space, as long as it lies entirely within a single isotherm plane. If plotted in a *PV* diagram, an isothermic process can appear as a line, a curve, a polygon, a circle, an ellipse—anything you can draw with a pencil while never lifting it from the graph. This variety is possible because we see the isotherm plane face-on, rather than edge-on as is the case with isobaric or isochoric processes.

The size of a balloon can be increased, but if enough air is added, the temperature inside does not change. The pressure inside a balloon can be reduced, but if the balloon shrinks to the right extent, again, the temperature stays the same. Anything can be done to the balloon or the gas inside, but as long as the temperature inside remains constant, the process is isothermic.

ADIABATIC PROCESSES

Finally, imagine some process that takes place, or is imposed upon, an enclosure in such a way that no heat transfer occurs between the gas inside and the external environment. That is to say, the gas in the enclosure neither gains nor loses thermal energy. This is an *adiabatic process*. In *PVT* space, such a process appears as a line or curve called an *adiabatic* that is not necessarily restricted to any isochor, isobar, or isotherm plane. When graphed in a *PV* diagram, an adiabatic can be mistaken for an isotherm unless we are aware of the conditions under which the process occurs.

Adiabatic and isothermic processes are not the same. Suppose that a thermally insulated chamber with rigid walls can be varied in size. If the chamber contains a fixed amount of gas and is suddenly made larger, the gas inside it drops in temperature, but no thermal energy is lost. This process is adiabatic but not isothermic. If, while the chamber is made larger, enough gas is added to keep the temperature constant, the added gas must transfer some thermal energy to the interior of the enclosure; otherwise the temperature would fall. That process is isothermic but not adiabatic.

FIRST LAW OF THERMODYNAMICS

In an enclosed system, thermal energy is conserved. The thermal energy added to such a system is equal to the sum of the work done by the system and the change in the thermal energy in the system. This is called the *first law of thermodynamics*.

Let E_{ti} be the thermal energy initially contained in a system such as a sample of confined dry gas. Let W represent the mechanical work done by (not on!) the system. Let E_{tf} be the thermal energy in the same system after a certain amount of thermal energy E_{t+} has been added. Then:

$$E_{t+} = W + (E_{tf} - E_{ti})$$

where all the quantities are in the same energy units, usually joules.

EFFICIENCY OF A CYCLIC HEAT ENGINE

An everyday example of a cyclic heat engine is an *internal-combustion engine* used in most automobiles and trucks. Any such machine goes through a *mechanical cycle* with which you are familiar if you've done work with motor vehicles. There is also a *thermal cycle*.

Let E_{ti} be the thermal energy initially contained in a cyclic heat engine. Let W represent the mechanical work done by the system. Then the efficiency *Eff* as a proportion or ratio is given by:

$$Eff = W / E_{ti}$$

The efficiency $Eff_\%$ as a percentage is:

$$Eff_\% = 100 \ W / E_{ti}$$

where W and E_{ti} are expressed in the same energy units, usually joules.

SECOND LAW OF THERMODYNAMICS

In real life, it is impossible to build a heat engine that converts all the initial thermal energy into mechanical work. This is an outgrowth of the *second law of thermodynamics*, which states, in its basic form, that thermal energy never flows *spontaneously* from a cooler environment or object to a warmer environment or object. Such a transfer of energy can be forced to take place—in a *heat pump* for example—but energy from some external source must always be expended in order to make it happen. You can think of this law as analogous to the rule that water never flows *spontaneously* uphill. (An external energy source such as an incoming tide or a mechanical pump can, of course, drive water uphill.)

Some heat engines are more efficient than others. But in any heat engine, no matter how well-engineered and carefully tuned, it is invariably the case that *Eff* < 1 and *Eff*$_\%$ < 100%. The difference between reality and perfection is manifest as *energy loss*. In an internal combustion engine, much of this loss appears in the *exhaust*. You can observe it as hot gas coming from the exhaust pipe of a car, gasoline-combustion generator, or other similar machine when it is running. Some energy is also lost because of friction among moving components.

Let E_{ti} be the thermal energy initially input to a heat engine. Let W represent the mechanical work done by the system. Let E_{tl} be the thermal energy loss. Then:

$$E_{ti} = W + E_{tl}$$

where all the quantities are in the same energy units, usually joules. This can be rearranged to obtain:

$$W = E_{ti} - E_{tl}$$

Substituting in the above formula for efficiency as a proportion, we get:

$$Eff = (E_{ti} - E_{tl}) / E_{ti}$$

and as a percentage:

$$Eff_\% = 100 \, (E_{ti} - E_{tl}) / E_{ti}$$

THE CARNOT ENGINE

A *Carnot engine* is a cyclic heat engine that obtains the largest possible amount of mechanical work from the available thermal energy. That is, *Eff* is maximized to the greatest practical extent.

Consider a Carnot engine that derives mechanical energy from a thermal energy source having an absolute temperature T_{hi}, expressed in kelvins. Suppose the engine operates against a *heat sink* (also called a *cold reservoir*) with an absolute temperature of T_{lo}, also expressed in kelvins. The gas in this engine goes through four well-defined steps or phases that compose the *Carnot cycle* as they occur in succession. At the start of the cycle, the gas has a certain volume and a certain temperature, shown by the point marked "Start" in Fig. 4-5.

- In the first phase shown by dashed curve A, the initial temperature of the gas is T_{hi}. The volume increases while the temperature remains constant. This is *isothermic expansion*. The pressure drops, and thermal energy is absorbed.

- In the second phase shown by dashed curve B, the gas volume continues to increase but no thermal energy is gained or lost. This is *adiabatic expansion*. The pressure continues to drop. The temperature goes down as well, reaching T_{lo} at the end of this phase.

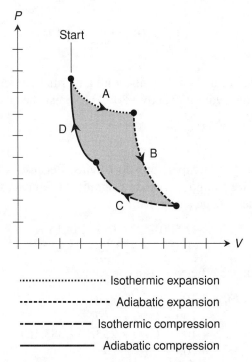

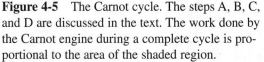

Figure 4-5   The Carnot cycle. The steps A, B, C, and D are discussed in the text. The work done by the Carnot engine during a complete cycle is proportional to the area of the shaded region.

- In the third phase shown by dashed curve C, the gas volume decreases while the temperature remains constant at T_{lo}. This is *isothermic compression*. The pressure rises. Thermal energy flows from the gas to the heat sink.

- In the fourth phase shown by solid curve D, the gas volume decreases but no thermal energy is gained or lost. This is *adiabatic compression*. The gas pressure rises. The temperature increases too, returning to T_{hi}. At the end of this phase, the gas temperature, volume, and pressure are all the same as they were at the beginning of the cycle.

- The amount of work done over the course of one complete cycle is proportional to the area enclosed by the four curves A, B, C, and D. In Fig. 4-5, this region is shaded.

The efficiency of a Carnot engine can be calculated in terms of the absolute temperatures T_{hi} and T_{lo}. As a proportion:

$$Eff = (T_{hi} - T_{lo}) / T_{hi}$$

As a percentage:

$$Eff_\% = 100 \, (T_{hi} - T_{lo}) / T_{hi}$$

PROBLEM 4-8

What is the efficiency of a Carnot engine with a thermal energy source at exactly +110°C and a cold reservoir at exactly –20°C? Express the answer to the nearest percentage point.

SOLUTION 4-8

First, we must convert the temperatures to kelvins. Because we are told the temperatures are exact, we can go to as many significant figures as we want. Adding 273.15 to both of these values gives us:

$$T_{lo} = -20.00 + 273.15 = 253.15 \text{ K}$$
$$T_{hi} = +110.00 + 273.15 = 383.15 \text{ K}$$

We can now calculate the efficiency as a percentage:

$$\begin{aligned} Eff_\% &= 100 \, (T_{hi} - T_{lo}) / T_{hi} \\ &= 100 - (100 \, T_{lo} / T_{hi}) \\ &= 100 - (100 \times 253.15 / 383.15) \\ &= 100 - 66 \\ &= 34\% \end{aligned}$$

PROBLEM 4-9

Under what conditions could the efficiency of a Carnot engine, in theory, be equal to 100%?

SOLUTION 4-9
Examine the formula for efficiency in terms of the absolute temperatures of the thermal energy source and the heat sink:

$$Eff_\% = 100 \, (T_{hi} - T_{lo}) / T_{hi}$$

From this, it is apparent that $Eff_\% = 100$ if and only if $T_{lo} = 0$ K and $T_{hi} > 0$ K. That means we could theoretically have a "perfect" heat engine if we could find a cold reservoir with a temperature of absolute zero. Such an ideal heat sink is not available to us, but we can come close to making one by cooling a large amount of helium gas until it liquefies. That happens at 4.2 K. It would also help if we were to employ a thermal energy source with the highest possible temperature.

Quiz

This is an "open book" quiz. You may refer to the text in this chapter. A good score is 8 correct. Answers are in the back of the book.

1. Imagine a sample of gas confined to a spherical enclosure whose diameter can be varied at will. Suppose the diameter of the sphere is suddenly doubled, but not a single molecule is allowed into or out of it. Also suppose that the termperature of the gas remains constant. What happens to the pressure of gas?

   (a) More information is necessary to answer this.
   (b) It becomes half as great.
   (c) It becomes 1/4 as great.
   (d) It becomes 1/8 as great.

2. Consider a simple experiment with a compressed-air canister designed for blowing dust from computer keyboards and electronic circuits. When you hold the button down for a while so a lot of air is released, the canister gets cold. Now suppose that you place the canister in a thermally insulating jacket that prevents heat transfer, and let some more air out. This process is

   (a) isochoric and adiabatic.
   (b) isothermic and adiabatic.
   (c) isobaric and isothermic.
   (d) isobaric and isochoric.

3. Suppose a Carnot engine operates according to the cycle shown in the *PV* diagram of Fig. 4-6. Four points on the cycle are shown as *W* (the start and end of the cycle) *X*, *Y*, and *Z*. Which of the following changes will most likely be accompanied by an increase in the work done by this system over the course of one complete cycle?

(a) Decreasing the pressure and increasing the volume slightly at point *W*, while leaving the pressure and volume at the other three points unchanged.

(b) Increasing the pressure and increasing the volume slightly at point *X*, while leaving the pressure and volume at the other three points unchanged.

(c) Increasing the pressure and decreasing the volume slightly at point *Y*, while leaving the pressure and volume at the other three points unchanged.

(d) Increasing the pressure and increasing the volume slightly at point *Z*, while leaving the pressure and volume at the other three points unchanged.

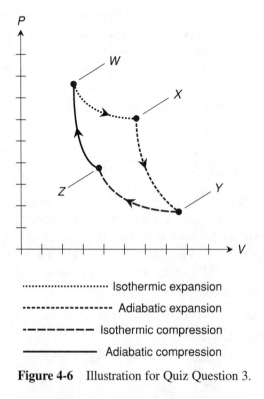

Figure 4-6 Illustration for Quiz Question 3.

4. What is the kelvin equivalent of –460°C, expressed to the nearest whole number?

(a) There is no such temperature.

(b) –187 K.

(c) –17 K.

(d) 0 K.

5. Imagine a solid sphere made of an unknown material with a surface area of 10.00 m^2 at 20°C. When the temperature increases to 30°C, the surface area of the sphere is 9.98 m^2. What is the thermal coefficient of *volume* expansion for this material, accurate to two significant figures?

(a) -9.0×10^{-4} /°C.

(b) -6.0×10^{-4} /°C.

(c) -3.0×10^{-4} /°C.

(d) It is impossible to calculate it without more information.

6. Refer to Problem and Solution 4-5 on page 116. Suppose that the neon gas is removed from the enclosure and pure, dry helium gas (which, like neon, is monatomic) is pumped in. Take the mass of a helium molecule to be 40.0 percent of the mass of a neon molecule. By what factor (if any) is the rms speed of the gas molecules in the enclosure different than it was before, assuming the temperature remains at exactly 0°C?

(a) The rms speed of the molecules is greater by a factor of 2.50.

(b) The rms speed of the molecules is greater by a factor of 1.58.

(c) The rms speed of the molecules is smaller by a factor of 2.50.

(d) The rms speed of the molecules is the same as it was before, because the temperature has not changed.

7. Refer to Problem and Solution 4-6 on page 116. Suppose that the neon gas is removed from the enclosure and pure, dry argon gas (which, like neon, is monatomic) is pumped in. Take the mass of an argon molecule to be 1.80 times the mass of a neon molecule. By what factor (if any) is the mean kinetic energy of the gas molecules in the enclosure different than it was before, assuming the temperature remains at exactly 0°C?

(a) The mean kinetic energy of the molecules is greater by a factor of 1.80.

(b) The mean kinetic energy of the molecules is greater by a factor of 1.34.

(c) The mean kinetic energy of the molecules is smaller by a factor of 1.34.

(d) The mean kinetic energy of the molecules is the same as it was before, because the temperature has not changed.

8. Suppose your home town is Kansas City, Missouri. You are visiting in Berlin, Germany on vacation, and you feel sick. You go to a doctor. She places a thermometer in your mouth. Then she takes it out and says, "Hmmm. Your temperature is 40 degrees, exactly." You are confused for a moment, and then you realize she is referring to your temperature in degrees Celsius. You pull your trusty calculator from your shirt pocket, perform the required operations, and conclude that

(a) your body temperature is normal.

(b) your body temperature is far above normal.

(c) your body temperature is a little below normal.

(d) your body temperature is far below normal.

9. The specific heat of pure liquid water, assuming no change of state occurs, can be expressed to four significant figures as

(a) 1.000 J/kg/K.

(b) 4.814 J/kg/K.

(c) 0.2390 J/kg/K.

(d) None of the above

10. Under what conditions can thermal energy be transferred to or from a material sample without causing a change in its temperature?

(a) Pure liquid water at sea level can absorb thermal energy while remaining at a temperature of 373.15 K.

(b) Water ice at sea level can absorb thermal energy while remaining at a temperature of −20°C.

(c) Pure water vapor at sea level can give up thermal energy while remaining at a temperature of +500°F.

(d) Forget it! Thermal energy can never be transferred to or from a material sample without causing a change in its temperature.

CHAPTER 5

Electrical and Magnetic Effects

Electrostatics is the physics of stationary, constant electric charges. *Magneto-statics* is the physics of stationary, constant magnetic pole pairs. In this chapter, we will study these phenomena along with the *electric fields* and *magnetic fields* they produce.

Electric Charge

Two types of electric charge exist. Scientists use the terms *positive* and *negative* (sometimes called *plus* and *minus*) to represent the two *charge polarities*.

REPULSION AND ATTRACTION

When two electrically charged objects are placed in close proximity, they either attract one another or repel one another. The force of attraction or repulsion, called the *electric force*, operates through space between any two electrically charged objects. Two objects attract each other if one has a positive charge and the other has a negative charge. If both objects are positively charged or both objects are negatively charged, they repel each other.

CHARGED SUBATOMIC PARTICLES

All matter is composed of *atoms* that are in turn made up of *protons*, *neutrons*, and *electrons*. Protons and neutrons tend to be lumped together at the cores of atoms. The core of an atom is called the *nucleus* (plural: *nuclei*). Electrons are much less massive than protons or neutrons, and they move around a lot more. Some electrons "orbit" a specific nucleus and stay there indefinitely. But it's common for an electron to travel from one atomic nucleus to another. Protons and neutrons, which compose the nucleus, don't move from atom to atom unless a *nuclear reaction* occurs.

Protons and electrons carry equal but opposite electric charges. By convention, protons are considered electrically positive, and electrons are considered electrically negative. The amount of charge on any proton is the same as the amount of charge on any other proton. Similarly, the amount of charge on any electron is the same as the amount of charge on any other electron. Although protons and electrons have opposite polarity, their net electric charge is the same in terms of quantity. A proton is "just as positive" as an electron is negative. Neutrons have no electrostatic charge.

A constant excess or deficiency of electrons on an object produces an electrostatic charge on that object. This is the usual way for a physical object to carry so-called "static electricity." If an object contains more total electrons than total protons, then that object is *negatively charged*. If an object contains fewer total electrons than total protons, then that object is *positively charged*.

UNITS OF CHARGE

The most straightforward way to measure electrostatic charge is to consider the charge on a single electron as the fundamental unit. This is known as an *electrostatic charge unit* (ecu), also called an *elementary charge unit*. Because the ecu is an extremely small unit of charge, a unit called the *coulomb* is more often used. One coulomb (1 C) constitutes more than six million-million-million elementary charge units! The relation between 1 C and 1 ecu, accurate to six significant figures according to NIST, can be expressed as follows:

$$1 \text{ C} = 6.24151 \times 10^{18} \text{ ecu}$$

or

$$1 \text{ ecu} = 1.60218 \times 10^{-19} \text{ C}$$

Electric Force

When two electrically charged objects are near each other and are not in relative motion, the magnitude and direction of the resulting force can be calculated based on the distance between the objects, the extent to which each object is charged, and the polarities of the charges.

COULOMB'S LAW

Imagine two electrically charged objects X and Y whose diameters are extremely small compared to the distance between them—so small that all their charge can be thought of as concentrated at specific geometric points. Such a charged object is known as a *point electric charge* or simply a *point charge*. (A small object with uniformly distributed charge can be considered as a point charge located at its geometric center.) Let q_x be the charge on object X, and let q_y be the electric charge on object Y. Let s be the distance between the charge centers. The repulsive force F_r between these two objects can be found by the following formula:

$$F_r = kq_x q_y / s^2$$

where k is a constant that depends on the units of force, charge, and distance we use, and on the nature of the medium separating the objects. In a perfect vacuum when F_r is expressed in newtons, q_x and q_y are expressed in coulombs, and s is expressed in meters, the value of k is approximately 8.98755×10^9 *newton meters squared per coulomb squared* ($N \cdot m^2/C^2$). This rule and the associated formula have become known as *Coulomb's law*. We can round off the value of k to $8.988 \times 10^9 \ N \cdot m^2/C^2$ for most purposes.

Coulomb's law does not hold perfectly if either or both of the charged objects have diameters that are more than a tiny fraction of the distance separating them. The law also fails in media other than a perfect vacuum. Dry air is almost like a vacuum as far as the value of k is concerned. However, other media, especially liquids and solids, can have values of k that are nowhere near $8.988 \times 10^9 \ N \cdot m^2/C^2$.

We specify F_r as a repulsive force. This allows us to put positive or negative charge values into our equations and always obtain the correct results for the direction of the force vector. If both q_x and q_y are positive, or if both are negative, then $F_r > 0$. That means the force acts as "positive repulsion" between X and Y, tending to drive them apart. If q_x is negative and q_y is positive or vice-versa, then $F_r < 0$, and the force is manifest as "negative repulsion"—that is, attraction—between X and Y, tending to draw them together.

THE LAW OF SUPERPOSITION

When a point charge X is near two or more other point charges, the net electric force vector on X, caused by all the other point charges collectively, is equal to the vector sum of the force vectors produced by each of the other point charges individually. This is called the *law of superposition* or the *superposition theorem* for electric force vectors.

Figure 5-1 illustrates the law of superposition in a simple case where a point charge X is influenced by two other point charges Y and Z. In this case X carries a positive charge, Y carries a negative charge, and Z carries a positive charge. In illustration A, vector $\mathbf{F}_{xy}$, representing the force produced on X by Y, is attractive, tending to pull X toward Y. Vector $\mathbf{F}_{xz}$, representing the force produced on X by Z, is repulsive, tending to push X away from Z. The actual force imposed on X by the combined electrostatic forces from Y and Z is shown as vector $\mathbf{F}_{xyz}$.

Vector $\mathbf{F}_{xyz}$ can be found in terms of $\mathbf{F}_{xy}$ and $\mathbf{F}_{xz}$ by vector addition, geometrically shown in Fig. 5-1B. To express this as a vector equation, we write:

$$\mathbf{F}_{xyz} = \mathbf{F}_{xy} + \mathbf{F}_{xz}$$

This formula can be extrapolated to any number of vectors in a sum, so we can determine the net electric force vector on a particular point charge no matter how

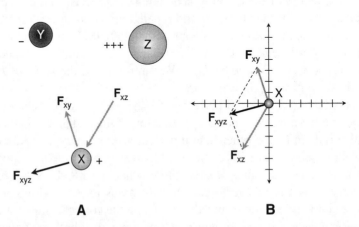

A   **B**

Figure 5-1 The law of superposition for electrostatic forces. Vector $\mathbf{F}_{xyz}$, the net force imposed on X by Y and Z, is equal to the sum of $\mathbf{F}_{xy}$, the force imposed on X by Y, and $\mathbf{F}_{xz}$, the force imposed on X by Z. Drawing A shows a hypothetical example. Drawing B shows the vector summation geometry for that example.

many other point charges there are in the vicinity, as long as we know the magnitudes and directions of the force vectors produced by each point charge.

Suppose we are given coordinates in a real-world system for a large number of point charges along with the actual charge quantity on each, and we want to find the net force vector imposed on any one point charge X by all the others. We can use Coulomb's law to find the electric force vectors on X caused by each of the other point charges. Then we can invoke the law of superposition, adding up all the individual force vectors. This can be a tedious process if the calculations are done "manually" when there are more than two particles involved, and especially if the system is three-dimensional! But a computer can be programmed to solve such a problem easily.

PROBLEM 5-1

Imagine two point charges in a vacuum. Suppose that the left-hand point charge carries 1 C of positive charge, and the right-hand point charge carries 1 C of negative charge (Fig. 5-2A). This results in an attractive force of magnitude F between the two objects. Now suppose that the charge quantity on each object is doubled, but the distance between their charge centers does not change. What happens to the force?

SOLUTION 5-1

The force is quadrupled to $4F$, and it remains an attractive force. The product of the charges increases by a factor of $2 \times 2 = 4$, as shown in Fig. 5-2B. This is true, however, only because the separation s between the point charges remains the same.

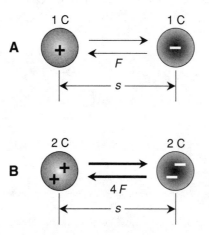

Figure 5-2   Illustration for Problem and Solution 5-1.

PROBLEM 5-2

Imagine two point charges in a vacuum. Suppose the left-hand point charge carries 1 C of positive charge, and the right-hand point charge carries 1 C of negative charge (Fig. 5-3A). This results in an attractive force of magnitude F between the two point charges. Suppose that the charge quantity on both of the objects is doubled, and the distance between their charge centers is also doubled. What happens to the force?

SOLUTION 5-2

The force does not change. Doubling the charge quantity on both objects but not changing the distance between their charge centers would increase the product of the charge quantities, and therefore the force, by a factor of 2×2 or 4. Increasing the distance between the charge centers from s to $2s$, without changing the charge quantity on either object, would cause the force to diminish by a factor equal to 2^2 or 4. In this situation both events take place, so the increase in force caused by the charge increase is exactly nullified by the decrease in force caused by the increase in the separation (Fig. 5-3B).

PROBLEM 5-3

Imagine three point charges X, Y, and Z in a vacuum. Suppose that X carries a charge quantity q_x of $+5.00 \times 10^{-5}$ C, Y carries a charge quantity q_y of $+2.00 \times 10^{-5}$ C, and Z carries a charge quantity q_z of -3.00×10^{-5} C. Further suppose that the point charges lie in a horizontal plane, and are arranged as shown in Fig. 5-4A. This is a system of *navigator's polar coordinates* where each angular division represents an *azimuth* (compass direction) increment of $10°$ and each radial division represents a *range* (radial displacement) increment of 10 cm. Draw the electrostatic force vector $\mathbf{F}_{xyz}$ exerted on X by the nearby point charges Y and Z.

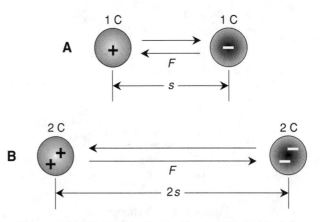

Figure 5-3   Illustration for Problem and Solution 5-2.

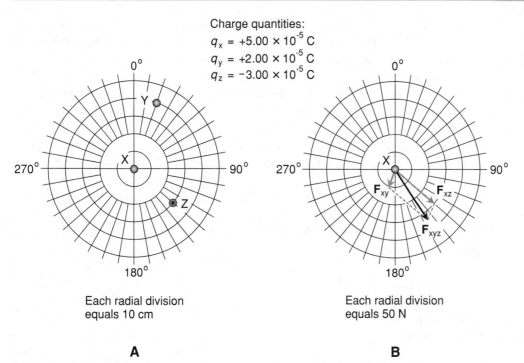

Charge quantities:
$$q_x = +5.00 \times 10^{-5}\,C$$
$$q_y = +2.00 \times 10^{-5}\,C$$
$$q_z = -3.00 \times 10^{-5}\,C$$

Each radial division
equals 10 cm

Each radial division
equals 50 N

A   **B**

Figure 5-4   Illustration for Problem and Solution 5-3.

SOLUTION 5-3

First, let's be sure we have our units correct. We're given all the charge quantities in coulombs, so they are all right. The radial displacement increments are given in centimeters, so we should convert these to meters. From this, it is evident that the distance s_{xy} between X and Y is 0.40 m, and the distance s_{xz} between Y and Z is 0.30 m. We can approximate the repulsive force magnitude F_{xy} between X and Y and the repulsive force magnitude F_{xz} between X and Z as follows:

$$
\begin{aligned}
F_{xy} &= (8.988 \times 10^9)\, q_x q_y \,/\, s_{xy}^{\,2} \\
&= 8.988 \times 10^9 \times 5.00 \times 10^{-5} \times 2.00 \times 10^{-5} \,/\, 0.40^2 \\
&= (8.988 \times 5.00 \times 2.00 \,/\, 0.16) \times 10^9 \times 10^{-5} \times 10^{-5} \\
&= 56\,N
\end{aligned}
$$

$$
\begin{aligned}
F_{xz} &= (8.988 \times 10^9)\, q_x q_z \,/\, s_{xz}^{\,2} \\
&= 8.988 \times 10^9 \times 5.00 \times 10^{-5} \times (-3.00) \times 10^{-5} \,/\, 0.30^2 \\
&= [8.988 \times 5.00 \times (-3.00) \,/\, 0.090] \times 10^9 \times 10^{-5} \times 10^{-5} \\
&= -150\,N
\end{aligned}
$$

The minus sign in the second calculation indicates that the force is attractive. Remember that the formula for Coulomb's law is for repulsive force magnitude. "Negative repulsion" is equivalent to attraction.

As we look at Fig. 5-4A, we can see that Y is at azimuth 20° relative to X. That means the force vector $\mathbf{F}_{xy}$, which repels X straight away from Y, points 180° opposite this, or toward azimuth 200°. We can also see that Z is at azimuth 130° relative to X. That means the force vector $\mathbf{F}_{xz}$, which attracts X straight toward Z, points toward azimuth 130°. We can plot $\mathbf{F}_{xy}$ and $\mathbf{F}_{xz}$ approximately as shown in Fig. 5-4B where each radial division represents 50 N. The vector sum, $\mathbf{F}_{xyz}$, can be drawn by constructing the diagonal of a parallelogram with $\mathbf{F}_{xy}$ and $\mathbf{F}_{xz}$ as adjacent sides. By examining the result in Fig. 5-4B, we can see that the magnitude of vector $\mathbf{F}_{xyz}$ is roughly 170 N, and the azimuth of vector $\mathbf{F}_{xyz}$ is roughly 147°. This graphical method of finding the net force vector is not very precise, but it gives us the general idea.

Electric Fields

An electrically charged particle or object is always surrounded by an *electric field*. This field has demonstrable effects on other charged particles or objects nearby. These effects are of interest not only to physicists, but to engineers, chemists, and even medical scientists.

LINES OF ELECTRIC FLUX

The electric field in the vicinity of a stationary point charge can be illustrated informally as *lines of electric flux*. These "lines" (which often turn out to be curves) are theoretical, not visible or material. Each "line" represents a certain quantity of electric flux. The magnitude of an electric field is defined in terms of *electric flux density* and is expressed in *coulombs per meter squared* (C/m^2). Sometimes the qualitative expressions "lines of flux per meter squared" or "flux lines per meter squared" are used to describe electric flux density.

A single positively charged object produces lines of electric flux that emanate straight outward from the charge center, as shown in Fig. 5-5A. A single negatively charged object produces lines of electric flux that converge straight inward toward the charge center, as shown in Fig. 5-5B. The direction of "flux flow" is away from positive charge poles and toward negative charge poles as defined by convention. The electric flux density in the vicinity of a point charge diminishes with increasing distance from the charge center according to the *inverse-square law*. For

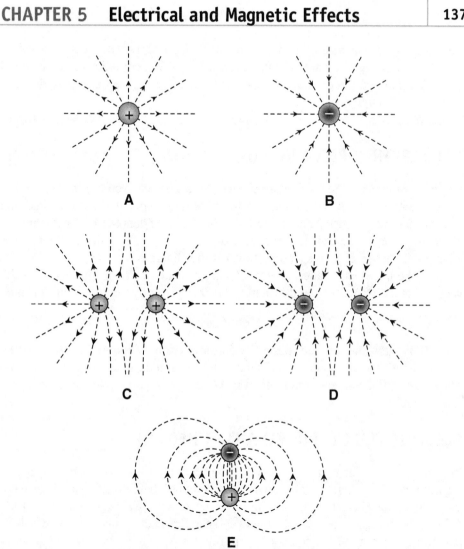

Figure 5-5   Electric flux lines in various situations. At A, an isolated, positively charged object. At B, an isolated, negatively charged object. At C, two
positively charged objects in proximity. At D, two negatively charged objects
in proximity. At E, two oppositely charged objects in proximity.

example, if the distance from a charge center is doubled, the electric flux density
is reduced to one-quarter of its previous value; if the distance increases by a factor of 10, the electric flux density becomes 1/100 as great.

When two positively charged objects are brought near each other, their electric
fields interact as shown in Fig. 5-5C, and repulsion occurs. When two negatively
charged objects are brought near each other, their electric fields interact as shown
in Fig. 5-5D, and again, repulsion occurs. If the charges are opposite, the electric

fields interact as shown in Fig. 5-5E, and attraction occurs. The electric flux density in the vicinity of a pair of point charges varies in a complicated manner and depends on the polarities and quantities of the charges, as well as on the distance between the charge centers.

THE ELECTRIC-FIELD VECTOR

Imagine a point charge in *free space* (ideally a perfect vacuum) with no other objects nearby. At any given location in space near a positive point charge, the *electric-field vector*, often called the *E-field vector* and denoted **E**, points directly away from the charge center, in the same direction as the "flux flow." Near a negative point charge, **E** points straight towards the charge center.

Let q be the electrostatic quantity of a point charge. The magnitude E of the vector **E** at a distance r from the charge center is given by the following formula:

$$E = kq \,/\, r^2$$

where k is the same constant described in the section "Coulomb's law," q is the charge quantity in coulombs, and r is the distance from the charge center in meters. The value of E is expressed in units of *newtons per coulomb* (N/C).

ELECTRIC FLUX THROUGH A FLAT REGION

Imagine a small, electrically non-conducting section S of a Euclidean (flat) plane. Suppose that S has an interior area of A_s. Also suppose that a constant, uniform electric-field vector **E** exists in the vicinity of S, such that dir **E** is perpendicular to S. This can be thought of as a set of straight and parallel electric flux lines, all of which pass through S at a 90° angle as shown in Fig. 5-6A. In this situation, the quantity of electric flux Φ_s that passes through S is equal to the product of E and A_s:

$$\Phi_s = EA_s$$

Electric flux quantity is expressed in *newton meters squared per coulomb* (N · m²/C).

Now suppose that dir **E** is not perpendicular to S, but instead subtends an angle θ with respect to a normal (perpendicular) line N that goes through S. This can be imagined as a set of straight and parallel electric flux lines, all of which pass through S at a slant (Fig. 5-6B). In a scenario like this, the total quantity of flux Φ_s that passes through S is equal to the product of E and A_s, times the cosine of the slant angle of the flux lines with respect to line N:

$$\Phi_s = EA_s \cos \theta$$

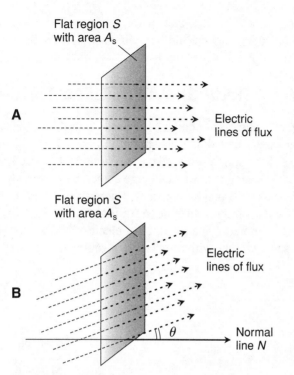

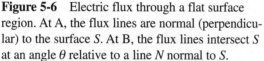

Figure 5-6   Electric flux through a flat surface region. At A, the flux lines are normal (perpendicular) to the surface *S*. At B, the flux lines intersect *S* at an angle θ relative to a line *N* normal to *S*.

In general, when we have a constant, uniform electric field in the vicinity of a flat region *S* with area A_s, the total flux Φ_s that passes through *S* is equal to the dot product of the electric-field vector **E** and a vector $\mathbf{A}_s$ whose magnitude is equal to the area A_s and whose direction is the same as that of the normal line *N*:

$$\Phi_s = \mathbf{E} \cdot \mathbf{A}_s$$

Sometimes the slant angle is defined between the lines of flux and the flat region *S* itself, rather than between the lines of flux and a line *N* normal to *S*. If we let the slant angle thus defined be symbolized ϕ, then:

$$\Phi_s = EA_s \cos(90° - \phi)$$

The preceding three formulas apply only in situations where the lines of flux are straight and parallel in the vicinity of the region through which they pass. If the lines of flux are not straight or not parallel, the situation becomes complicated. We won't

worry about situations like that, except to consider what happens when a point charge is surrounded by a sphere through which straight, radial lines of flux pass.

ELECTRIC FLUX THROUGH A SPHERICAL REGION

Imagine an electrostatic point charge of quantity q at the center of an electrically non-conducting sphere of radius r, as shown in Fig. 5-7. Consider a small region S on the surface of the sphere, such that S has area A_s. Electric lines of flux, emanating from or converging toward the charge center, pass through S. At every point on S, the flux lines are perpendicular to S. When we look at S close-up, the situation is essentially the same as that shown in Fig. 5-6A. (The slight spherical curvature of S can be ignored.) The quantity of electric flux Φ_s that passes through S is found in the same way as with a flat surface region:

$$\Phi_s = EA_s$$

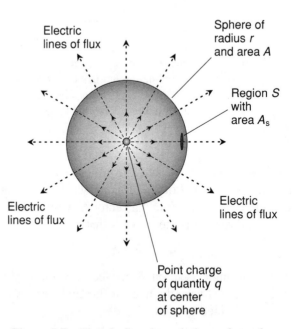

Figure 5-7 Electric flux through the surface of a sphere with a point charge at its center. The total flux passing through the sphere does not depend on the size of the sphere, but only on the charge quantity inside.

Now consider the ratio of the area A of the whole sphere to the area A_s of region S. This ratio is an expression of the number of regions identical to S that can be figuratively "pasted onto" the sphere to cover its surface without any gaps or overlap. The total flux quantity Φ that passes through the sphere is equal to the flux passing through S times the ratio of A to A_s, as follows:

$$\Phi = \Phi_s A / A_s$$

Rearranging this formula to get Φ_s in terms of the other quantities, we obtain:

$$\Phi_s = \Phi A_s / A$$

By substitution in the formula $\Phi_s = EA_s$ from above:

$$EA_s = \Phi A_s / A$$

Transposing the left-hand and right-hand sides of this equation:

$$\Phi A_s / A = EA_s$$

Multiplying each side through by A / A_s and then simplifying gives us this:

$$\Phi = AE$$

We are told that the sphere has radius r. Recall the formula for the surface area A of a sphere in terms of its radius r:

$$A = 4\pi r^2$$

By substitution we obtain:

$$\Phi = 4\pi r^2 E$$

Now remember the formula for E in terms of charge quantity:

$$E = kq / r^2$$

Again we can substitute and simplify, obtaining:

$$\Phi = 4\pi r^2 \, (kq / r^2)$$
$$= 4\pi kq$$

where k is that familiar constant whose value is approximately $8.988 \times 10^9 \, \mathrm{N \cdot m^2/C^2}$.

Note that the total flux quantity Φ through a sphere with a point charge of quantity q at its center does not depend on the size of the sphere. If a tiny point charge is centered inside a sphere the size of a basketball or the size the moon, the total flux quantity passing through that sphere is the same in either case! The flux density Φ_s through any small region S of constant area on the surface of the sphere, however, does depend on the size of the sphere, being inversely proportional to the square of the sphere's radius.

PERMITTIVITY

As things work out, the quantity $1 / (4\pi k)$, where k is the constant defined above, is equal to a universal physical constant called the *permittivity of free space* or the *electric constant*. In simplistic terms, *permittivity* is an expression of the extent to which a medium can store energy in the form of an electric field. The permittivity of free space is symbolized by the lowercase Greek letter epsilon with a subscript zero (ε_0). As given by NIST to six significant figures, its value is:

$$\varepsilon_0 = 8.85419 \times 10^{-12}$$

expressed in *coulombs squared per newton meter squared*, symbolized $C^2/(N \cdot m^2)$. It is often rounded to $8.855 \times 10^{-12} \, C^2/(N \cdot m^2)$. The relation between ε_0 and k can be expressed as follows:

$$\varepsilon_0 = 1 / (4\pi k)$$
$$\text{or}$$
$$4\pi k = 1 / \varepsilon_0$$

By substitution from above:

$$\Phi = 4\pi k q$$
$$= q / \varepsilon_0$$

We can rearrange this formula to obtain:

$$q = \Phi \varepsilon_0$$

By substitution in the earlier formula for the magnitude E of the vector $\mathbf{E}$ on the surface of a sphere of radius r centered on a point charge of quantity q, we have:

$$E = kq / r^2$$
$$= k\Phi \varepsilon_0 / r^2$$

Recall that $4\pi k = 1 / \varepsilon_0$. We can rearrange this to obtain:

$$k = 1 / (4\pi \varepsilon_0)$$

Then substituting again:

$$E = k\Phi \varepsilon_0 / r^2$$
$$= [1 / (4\pi \varepsilon_0)] \, \Phi \varepsilon_0 / r^2$$
$$= \Phi / (4\pi r^2)$$

This gives us a formula for determining E at a distance r from a point charge when only the total electric flux surrounding that point charge is known. The quantity $4\pi r^2$ is equal to the surface area A of a sphere of radius r. Substituting yet again:

$$E = \Phi / A$$

There are two other ways of expressing this same relation, one of which we have already stated:

$$\Phi = AE$$
$$A = \Phi / E$$

This tells us that the total flux quantity Φ through a sphere with a point charge of quantity q at its center does not depend on the permittivity of the medium inside the sphere. Nevertheless, the flux density Φ_s through any given small region S on the surface may vary from place to place if the permittivity of the medium is not uniform throughout the interior of the sphere.

GAUSS'S LAW

The total quantity of electric flux that passes through any closed, non-conducting surface surrounding a point charge is always equal to the charge quantity divided by the permittivity of free space. This is true whether or not the medium in the enclosure is actually free space, and whether or not the enclosure is a sphere. It is also true whether the point charge is positive (producing an "outward flux flow") or (negative producing an "inward flux flow"). These claims are based on four facts in spatial topology:

- To get from the inside of any closed surface to the outside, you must pass through the surface at some point exactly once. (This applies to a positive point charge.)
- To get from the outside of any closed surface to the inside, you must pass through the surface at some point exactly once. (This applies to a negative point charge.)
- If you are inside a closed surface and then pass through it at some point exactly once, you end up outside. (This applies to a positive point charge.)
- If you are outside a closed surface and then pass through it at some point exactly once, you end up inside. (This applies to a negative point charge.)

Now imagine that there are n point charges inside a closed surface. Suppose that the objects have charge quantities q_1, q_2, q_3, ..., and q_n. If the total flux values caused by charges q_1, q_2, q_3, ..., and q_n are Φ_1, Φ_2, Φ_3, ..., and Φ_n respectively, then the net flux Φ passing through the closed surface as a result of the combined effects of all the charges can be calculated as follows:

$$\Phi = \Phi_1 + \Phi_2 + \Phi_3 + \ldots + \Phi_n$$
$$= q_1 / \varepsilon_o + q_2 / \varepsilon_o + q_3 / \varepsilon_o + \ldots + q_n / \varepsilon_o$$
$$= (q_1 + q_2 + q_3 + \ldots q_n) / \varepsilon_o$$

This rule was first rigorously demonstrated by the mathematician, astronomer, and physicist Carl Friedrich Gauss (1777–1855) and has become known as *Gauss's law*.

PROBLEM 5-4

Prove that the flux quantity passing at right angles through a region of fixed area near a point charge is inversely proportional to the square of the distance from the charge center. Assume that the medium has uniform permittivity, and that there are no other charged objects in the vicinity.

SOLUTION 5-4

Consider a sphere of radius r_1, and a small region S on its surface with area A_s as shown in Fig. 5-8A. Now imagine a second sphere of radius r_2, and a small region S on its surface with the same area A_s, as shown in Fig. 5-8B. Suppose that at the center of each sphere, there is a point charge having charge quantity q. The total

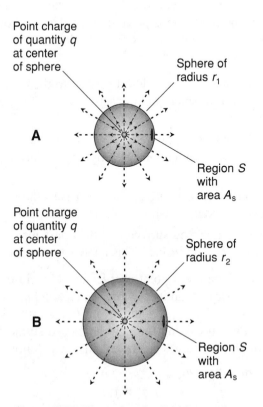

Figure 5-8   Illustration for Problem and Solution 5-4.

flux Φ passing through either sphere is therefore equal to q / ε_0. (Remember that the total flux depends only on the charge quantity and the permittivity of free space, and not on the size of the sphere or the actual permittivity of the medium inside.) All the flux lines pass through the regions S at right angles because all the flux lines are radial, the regions are on the sphere surfaces, and the charge centers are at the sphere centers.

In the situation of Fig. 5-8A, the ratio X_1 of the area of S to the area of the whole sphere is:

$$X_1 = A_s / (4\pi r_1^2)$$

The flux quantity Φ_1 through S in Fig. 5-8A is therefore equal to the ratio X_1 multiplied by the flux quantity Φ through the whole sphere:

$$\Phi_1 = X_1 \, \Phi$$

Now suppose that the distance of S from the charge center in Fig. 5-8B is k times the distance of S from the charge center in Fig. 5-8A. This means $r_2 = kr_1$. Then in Fig. 5-8B, the ratio X_2 of the area of S to the area of the whole sphere is:

$$\begin{aligned}
X_2 &= A_s / (4\pi r_2^2) \\
&= A_s / [4\pi(kr_1)^2] \\
&= A_s / (4\pi r_1^2 k^2) \\
&= (1 / k^2) [A_s / (4\pi r_1^2)] \\
&= (1 / k^2) X_1
\end{aligned}$$

The flux quantity Φ_2 through S in Fig. 5-8B is therefore equal to the ratio X_2 multiplied by the flux quantity Φ through the whole sphere:

$$\begin{aligned}
\Phi_2 &= X_2 \, \Phi \\
&= (1 / k^2) X_1 \, \Phi \\
&= (1 / k^2) \Phi_1
\end{aligned}$$

If the distance from a charge center changes by a factor of k, the resulting flux quantity passing at right angles through a region of fixed area changes by a factor of $1 / k^2$, or inversely according to the square of k. This is the statement we intended to prove.

PROBLEM 5-5

Suppose a completely closed, large, electrically non-conducting cube contains five point charges, all clumped close to the cube center, having the following values:

$$\begin{aligned}
q_1 &= +5.37 \times 10^{-6} \text{ C} \\
q_2 &= -1.05 \times 10^{-6} \text{ C} \\
q_3 &= +8.98 \times 10^{-6} \text{ C} \\
q_4 &= -7.77 \times 10^{-6} \text{ C} \\
q_5 &= -9.21 \times 10^{-6} \text{ C}
\end{aligned}$$

What is the total flux through this cube? Express the answer to three significant figures. What is the general direction of the E-field vector $\mathbf{E}$ at any point on the cube surface? Assume that the cube contains nothing other than the point charges. What will happen if one of the positive point charges is moved away from the cube center and placed very close to the cube surface (but still inside the cube)?

SOLUTION 5-5
First, let's find the sum of the charge quantities, and call this net charge q, as follows:

$$q = [(+5.37) + (-1.05) + (+8.98) + (-7.77) + (-9.21)] \times 10^{-6}$$
$$= -3.68 \times 10^{-6} \text{ C}$$

The total flux Φ through the cube is:

$$\Phi = q / \varepsilon_o$$
$$= (-3.68 \times 10^{-6}) / (8.855 \times 10^{-12})$$
$$= -4.16 \times 10^5 \text{ N} \cdot \text{m}^2/\text{C}$$

The minus sign in this result indicates that the net charge inside the cube is negative, so the flux "flows" towards toward the cube center. At any point on the cube surface, therefore, dir $\mathbf{E}$ is generally outside-in. This is the case because the point charges are all near the cube center. If one of the positive point charges is placed very close to the cube surface (but still inside the cube), then dir $\mathbf{E}$ in its immediate vicinity will be inside-out, even though the average "flux flow" from all the point charges taken together will be outside-in. The net flux quantity through the cube surface will be $-4.16 \times 10^5 \text{ N} \cdot \text{m}^2/\text{C}$ in any case, as long as all the point charges are inside the cube.

Magnetic Poles

Magnetism exists whenever electric *charge carriers* move relative to other objects, or relative to a frame of reference. Moving charge carriers produce *magnetic fields* and consequent *magnetic forces*.

LINES OF MAGNETIC FLUX

Physicists consider magnetic fields to be made up of *flux lines*, or *lines of magnetic flux*. The intensity of a magnetic field is determined according to the number of flux lines passing at a right angle through a region having a certain cross-sectional area, usually 1 centimeter squared (1 cm^2) or 1 meter squared (1 m^2). The lines are imaginary, not solid threads or fibers. But they are a useful concept in defining the geometry of a magnetic field.

NORTH AND SOUTH POLES

A magnetic field has a direction, or orientation, at any point in space near a current-carrying wire or a permanent magnet. The flux lines run parallel with the direction of the *magnetic field vector*, also called the *B-field vector* and symbolized **B**, at every point in space. A magnetic field is defined by convention to originate at a *north pole* and to terminate at a *south pole*. Usually, the lines (or curves) of flux diverge from north poles and converge toward south poles. Figure 5-9 shows the general shape of the flux field in the vicinity of two magnetic north poles close together (A), two magnetic south poles close together (B), and a south pole close to a north pole (C).

The north and south poles of a permanent magnet are not the same as the north and south magnetic poles of the earth, which are called the *geomagnetic poles*. The *north geomagnetic pole* is actually a south magnetic pole; it attracts the north poles

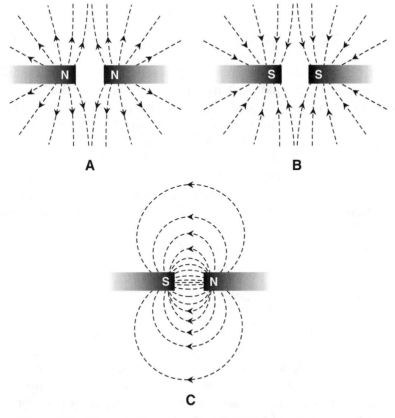

Figure 5-9 Magnetic flux geometry in situations involving two poles near each other. At A, two north poles in close proximity. At B, two south poles in close proximity. At C, a south pole near a north pole.

of permanent or artificial magnets. The *south geomagnetic pole* is actually a north magnetic pole; it attracts the south poles of permanent or artificial magnets.

REPULSION AND ATTRACTION

When magnetic poles are placed in close proximity, they either attract or repel, in a manner analogous to the behavior of electric poles. The magnetic force can operate through empty space or through any medium that does not "stick to a magnet."

Two magnetic south poles or two magnetic north poles produce a repulsive force when they are brought close together. When a north pole and a south pole are placed in proximity, attraction occurs. Whether repulsive or attractive, the force increases as the separation between the poles decreases. Also, the magnitude of the force, given a constant pole separation, depends on the "strength" of the magnetic fields. We'll expand on the concept of magnetic field strength shortly.

MONOPOLES AND DIPOLES

A charged electric particle, hovering in space and not moving, constitutes an *electric monopole*. That means it has only one pole. The flux lines around an electric monopole aren't closed. A positive charge does not have to be mated with a negative charge, although it can be. When there is a positive electric pole near a negative pole, the pair forms an *electric dipole*.

A *magnetic monopole*, in contrast to an electric monopole, can't ordinarily exist. North and south magnetic poles always occur together, except in certain peculiar human-made scenarios. Therefore, a magnetic field is always associated with a *magnetic dipole* (north-south pole pair). The lines of flux in the vicinity of a magnetic dipole always connect the two poles. Some flux lines are straight in a local sense, but in the larger sense they are curves.

A CURRENT-CARRYING WIRE

In the magnetic field around a straight wire carrying a constant electric current, the lines of flux are circular in any plane perpendicular to the wire (Fig. 5-10). The center of each flux circle lies at the point where the plane intersects the wire. These flux circles don't originate or terminate anywhere, so it is hard to imagine how such a magnetic field can have poles. Nevertheless, there is a way to define magnetic poles in the vicinity of a current-carrying wire.

Imagine two distinct points on one of the circular magnetic lines of flux in a plane perpendicular to the wire. A magnetic dipole is formed by the flux going

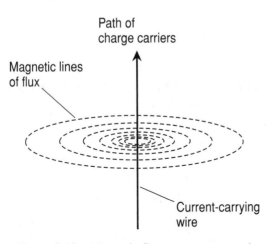

Figure 5-10   Magnetic flux geometry around a straight wire carrying constant current.

from one of these points to the other. This can be called a "virtual magnet" because it is a theoretical line segment connecting two points in space (Fig. 5-11). The flux moves out from the north pole and goes toward the south pole. In this drawing, the *conventional current*, which flows from electrical plus to minus, is from left to right. The electrons in the wire move from right to left.

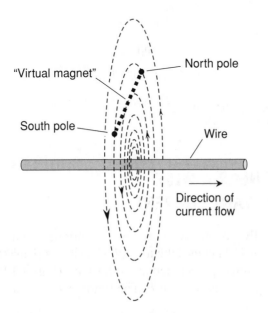

Figure 5-11   A method of defining magnetic poles near a straight wire carrying constant electric current.

Magnetic Flux

The overall size or extent of a magnetic field, called the *magnetic flux quantity*, is expressed in units called *webers* (Wb). A smaller unit, the *maxwell* (Mx), is sometimes used if the magnetic flux quantity is small. The weber and the maxwell are related as follows:

$$1 \text{ Wb} = 10^8 \text{ Mx}$$

and

$$1 \text{ Mx} = 10^{-8} \text{ Wb}$$

UNITS OF FLUX DENSITY

The *magnetic flux density*, or number of "flux lines per unit cross-sectional area," is a more useful expression for magnetic effects than the overall quantity of magnetism. Flux density is customarily denoted as the uppercase, italic letter B in equations.

A flux density of one *tesla* (1 T) is equal to one weber per meter squared (1 Wb/m^2). A flux density of one *gauss* (1 G) is equal to one maxwell per centimeter squared (1 Mx/cm^2). The tesla and the gauss are related like this:

$$1 \text{ T} = 10^4 \text{ G}$$

and

$$1 \text{ G} = 10^{-4} \text{ T}$$

For these definitions of flux density to be valid, the lines of flux must all pass through a surface at right angles.

FLUX DENSITY NEAR A WIRE CARRYING CONSTANT CURRENT

Imagine an electrically conducting wire that is infinitely thin and absolutely straight. Suppose it carries a constant current I. Consider a point P at a distance r from the wire as measured perpendicular to the wire (Fig. 5-12). Let B be the magnetic flux density in teslas at point P. The following formula applies:

$$B = 2 \times 10^{-7} \, (I \, / \, r)$$

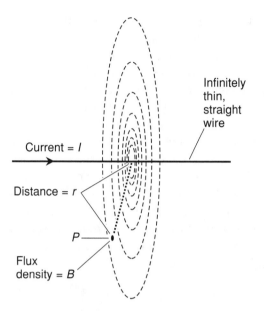

Figure 5-12   The flux density B at a point P near a current-carrying wire varies with the current I in the wire, and also with the distance r from the wire.

where I is in amperes and r is in meters. In this equation, the constant 2×10^{-7} can be considered exact to any required number of significant figures. If you want to calculate the magnetic flux density B in gauss rather than in teslas, use this formula:

$$B = 2 \times 10^{-3} \, (I \, / \, r)$$

where, again, I is in amperes and r is in meters. Here, the constant 2×10^{-3} can be considered exact to any required number of significant figures.

As long as the thickness of the wire is small compared with the distance r from it, and as long as the wire is reasonably straight in the vicinity of the point P at which the flux density is measured, these formulas provide a good representation of what happens around current-carrying electrical conductors in the real world.

AMPERE'S LAW

Remember, physicists define conventional current as flowing from the positive electric pole to the negative electric pole (plus to minus). This is opposite from the direction that electrons move. Suppose the conventional current portrayed in Fig. 5-13 flows out of the page toward you. This means that the region above the page is electrically negative relative to the region below the page. According to a rule called *Ampere's law*, the direction of the magnetic flux is counterclockwise in this situation. When conventional current flows toward you, the resulting magnetic flux turns counterclockwise from your point of view.

Ampere's law is sometimes called the *right-hand rule for magnetic flux*. If you hold your right hand with the thumb pointing out straight and the fingers curled, and then point your thumb in the direction of the conventional current flow in a straight wire, your fingers curl in the sense of the magnetic flux rotation. Similarly, if you orient your hand so that your fingers curl in the sense of the magnetic flux rotation and then straighten out your thumb, your thumb points in the direction of the conventional current. If you want to determine the sense of the magnetic flux rotation relative to the flow of electrons in the wire (that is, from minus to plus), use your left hand instead of your right hand.

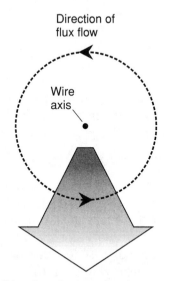

Figure 5-13   Ampere's law defines the direction in which magnetic flux "flows" around a straight wire carrying constant current.

PERMEABILITY

Do you wonder where the constants 2×10^{-7} and 2×10^{-3} come from in the above equations for flux density near a straight wire carrying constant current? These constants arise from a well-known constant called the *permeability of free space* divided by 2π, the number of radii in a perfect circle. In simplistic terms, *permeability* is an expression of the extent to which a medium can store energy in the form of a magnetic field. Permeability is the magnetic counterpart of electric permittivity.

The permeability of free space, also called the *magnetic constant*, is symbolized by the lowercase italic Greek letter mu with a subscript zero (μ_o). Its value is $4\pi \times 10^{-7}$ when expressed in *tesla meters per ampere* (T · m/A), or $4\pi \times 10^{-3}$ when expressed in *gauss meters per ampere* (G · m/A). As given by NIST to six significant figures, these values are:

$$\mu_o = 1.25664 \times 10^{-6} \text{ T} \cdot \text{m/A}$$
$$= 0.0125664 \text{ G} \cdot \text{m/A}$$

When we divide these constants by the number of radii in a circle (the shape of any flux curve surrounding a straight wire in free space carrying constant current), we get precisely 2×10^{-7} and 2×10^{-3}, the constants in the equations for flux density near a straight wire. Thus, the formulas become:

$$B_T = (\mu_o / 2\pi) \times 10^{-7} \, (I / r)$$

and

$$B_G = (\mu_o / 2\pi) \times 10^{-3} \, (I / r)$$

for the magnetic flux density B_T in teslas or B_G in gauss at a point near a current-carrying wire in free space, where I is in amperes and r is in meters.

Some *ferromagnetic materials* have permeability values larger than the permeability of free space. These media tend to concentrate magnetic lines of flux. Certain metal alloys have values of μ that are hundreds (or even thousands) of times μ_o, and are used in components known as *inductors* that store energy as a magnetic field. These alloys can also serve as *cores* for powerful *electromagnets*. A few substances have values of μ less than μ_o (but never very much less), and are called *diamagnetic materials*.

PROBLEM 5-6

What is the flux density in teslas at a distance of 255 millimeters (mm) from a straight, thin wire carrying 165 milliamperes (mA) of direct current? Express the answer to three significant figures.

SOLUTION 5-6

First, convert distance to meters and current to amperes. Thus $r = 0.255$ m and $I = 0.165$ A. Then plug these numbers into the formula. If B_T is the flux density in teslas, we have:

$$B_T = 2.00 \times 10^{-7} \, (I / r)$$
$$= 2.00 \times 10^{-7} \, (0.255 / 0.165)$$
$$= 3.09 \times 10^{-7} \text{ T}$$

PROBLEM 5-7

In the above scenario, what is the flux density in *milligauss* (thousandths of a gauss) at point P? Express the answer to three significant figures.

SOLUTION 5-7

To solve this, let's convert the answer from the previous problem from teslas to gauss, and then convert the result to milligauss. If B_G is the flux density in gauss, then:

$$B_G = 3.09 \times 10^{-7} \times 10^4$$
$$= 3.09 \times 10^{-3} \text{ G}$$
$$= 0.00309 \text{ G}$$

To convert this to milligauss, we must multiply (not divide!) by 1000, because there are 1000 milligauss per gauss. If B_{mG} is the flux density in milligauss, then:

$$B_{mG} = 0.00309 \times 1000$$
$$= 3.09 \text{ mG}$$

Magnetic Interaction

When magnetic fields interact with charged particles or current-carrying conductors, the magnitude and direction of the resulting *magnetic force* can be calculated if we have certain information about the interacting media.

MAGNETIC FORCE ON POINT CHARGE MOVING AT CONSTANT VELOCITY

Imagine a point charge P that moves at a constant velocity $\mathbf{v}_p$ in a magnetic field $\mathbf{B}$ that has straight, parallel lines of flux, as shown in Fig. 5-14. Let q_p be the charge

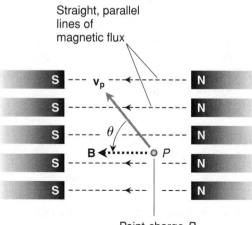

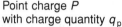

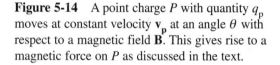

Figure 5-14   A point charge P with quantity q_p moves at constant velocity $\mathbf{v_p}$ at an angle θ with respect to a magnetic field $\mathbf{B}$. This gives rise to a magnetic force on P as discussed in the text.

quantity on P. The motion of P produces a magnetic field that interacts with $\mathbf{B}$ to produce a magnetic force $\mathbf{F_p}$ on P. The vector equation for determining $\mathbf{F_p}$ is:

$$\mathbf{F_p} = q_p\,(\mathbf{v_p} \times \mathbf{B})$$

We can express this in scalar form if we let F_p be the magnitude of the force vector $\mathbf{F_p}$, v_p be the magnitude of the velocity vector $\mathbf{v_p}$, and B be the magnitude of the magnetic-field vector $\mathbf{B}$, as follows:

$$F_p = |q_p|v_p B \sin \theta$$

Here, F_p is in newtons, q_p is in coulombs, v_p is in meters per second, B is in teslas, and θ is the angle between $\mathbf{v_p}$ and $\mathbf{B}$ expressed in the rotational sense from $\mathbf{v_p}$ toward $\mathbf{B}$, oriented so the extent of the rotation is between $0°$ and $180°$.

Note that the charge quantity q_p can be either positive or negative. This has an effect on dir $\mathbf{F_p}$. In the scenario of Fig. 5-14, the cross product of the particle-velocity and magnetic-field vectors, $\mathbf{v_p} \times \mathbf{B}$, points out of the page at a right angle. If the charge q_p is positive, then $\mathbf{F_p}$ points out of the page towards you. If the charge q_p is negative, then $\mathbf{F_p}$ points in exactly the opposite direction. In the scalar formula above, we specify the absolute value of q_p in order to avoid ever getting a "negative force magnitude."

In general, if all other factors are held constant, the magnitude of the magnetic force on a moving, charged particle in a magnetic field is greatest when the particle moves at a right angle to the magnetic lines of flux. As the angle between the particle path and the lines of flux decreases toward zero or increases toward 180°, the force magnitude decreases until, if the particle moves parallel to the lines of flux, there is no magnetic force on it.

MAGNETIC FORCE ON A LENGTH OF WIRE CARRYING CONSTANT CURRENT

The above-described situation can be expanded upon to evaluate the magnetic force on a straight electrical conductor that carries a constant current in a magnetic field. Imagine a thin, straight wire S that carries a constant current I_s over a displacement vector $\mathbf{s}$ in a magnetic field $\mathbf{B}$ that has straight, parallel lines of flux, as shown in Fig. 5-15. The motion of each electron in S produces a magnetic field around S; the sum of all these tiny fields causes the current I_s to generate its own magnetic field. This field interacts with $\mathbf{B}$ to produce a magnetic force vector $\mathbf{F}_s$ on the length of S defined by $\mathbf{s}$, according to the following formula:

$$\mathbf{F}_s = I_s \,(\mathbf{s} \times \mathbf{B})$$

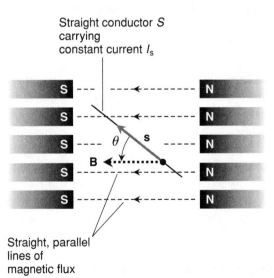

Straight conductor S
carrying
constant current I_s

Straight, parallel
lines of
magnetic flux

Figure 5-15   A straight electrical conductor S carrying current I_s over a displacement $\mathbf{s}$ at an angle θ with respect to a magnetic field $\mathbf{B}$. This gives rise to a magnetic force on S as discussed in the text.

We can express this in scalar form if we let F_s be the magnitude of $\mathbf{F_s}$, s be the magnitude of $\mathbf{s}$, and B be the magnitude of $\mathbf{B}$, as follows:

$$F = I_s \, sB \sin \theta$$

In this formula F_s is in newtons, I_s is in amperes, s is in meters, B is in teslas, and θ is the angle between $\mathbf{s}$ and $\mathbf{B}$ expressed in the rotational sense from $\mathbf{s}$ toward $\mathbf{B}$, oriented so the extent of the rotation is between $0°$ and $180°$.

In calculations such as these, current must be expressed in conventional form as a movement of charge carriers from the more positive electric pole to the more negative electric pole. The flow of electrons is always in the opposite direction from the direction of conventional current. In the scenario of Fig. 5-14, the cross product of the displacement and magnetic-field vectors, $\mathbf{s} \times \mathbf{B}$, and therefore $\mathbf{F_s}$, points at a right angle out of the front of the page, toward you. Therefore, if the charge carriers are electrons as they usually are in a metal wire, $\mathbf{F_s}$ points at a right angle out of the back of the page, away from you.

If all other factors are held constant, the magnitude of the magnetic force on a length of current-carrying conductor is greatest when the conductor is perpendicular to the magnetic lines of flux. As the angle between $\mathbf{s}$ and the lines of flux decreases toward zero or increases toward $180°$, the force magnitude decreases until, if the length of conductor defined by $\mathbf{s}$ is parallel to the lines of flux, $\mathbf{B}$ gives rise to no magnetic force on that length of wire, regardless of how much current the wire carries.

PROBLEM 5-8

Figure 5-16 is a simplified cross-sectional drawing of an *electromagnetic cathode-ray tube* (CRT). An electron "gun" emits electrons that travel from left to right through a series of accelerating electrodes called *anodes*. Then the electrons pass at a right angle across magnetic lines of flux generated by a pair of electromagnets. The magnetic field causes the electron beam to deflect because a force is exerted on each electron as it cuts through the flux. In the scenario shown here, let the velocity vector of each electron, as it approaches the magnetic field, be symbolized $\mathbf{v_e}$, and let the magnetic field vector between the electromagnets be symbolized $\mathbf{B}$. In what direction is the electron beam bent as it passes through the magnetic field if the electron speed v_e is a constant 1.5 km/s and the magnetic-field strength B is a constant 50 G? Consider the charge on a single electron to be $q_e = -1.00 \text{ ecu} = -1.60 \times 10^{-19} \text{ C}$. What is the magnitude F_e of the magnetic force vector $\mathbf{F_e}$ on a single electron? Express the answer to two significant figures.

SOLUTION 5-8

Note that dir $\mathbf{v_e}$ is from left to right, and dir $\mathbf{B}$ is straight down (from magnetic north to magnetic south) as shown in Fig. 5-8. According to the right-hand rule for

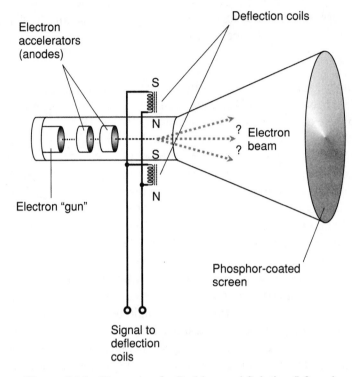

Electron
accelerators
(anodes)

Deflection coils

S

N

S

N

? Electron
? beam

Electron "gun"

Phosphor-coated
screen

Signal to
deflection
coils

Figure 5-16   Illustration for Problem and Solution 5-8, and
for Quiz Question 2.

cross products, dir ($\mathbf{v_e} \times \mathbf{B}$) is directly away from us, out of the back of the page.
Because electrons carry negative charge, dir $\mathbf{F_e}$ is opposite dir ($\mathbf{v_e} \times \mathbf{B}$), or straight
out of the page at us. That means the electron beam is deflected toward us. The
magnitude of the force vector on a single electron can be found as follows:

$$F_e = |q_e| v_e B \sin \theta$$

We can set $\theta = 90°$ because the electrons cut across the lines of flux at a right
angle. Therefore, $\sin \theta = 1$. Converting the magnetic field strength from gauss to
teslas, we have $B = 5.0 \times 10^{-3}$ T. Converting the electron speed from kilometers
per second to meters per second, we have $v_e = 1.5 \times 10^3$ m/s. Now we can calculate:

$$F_e = |q_e| v_e B$$
$$= 1.60 \times 10^{-19} \times 1.5 \times 10^3 \times 5.0 \times 10^{-3}$$
$$= 1.2 \times 10^{-18} \text{ N}$$

PROBLEM 5-9

Imagine a straight wire that carries a constant current of I_s = 85 mA, oriented as shown in Fig. 5-17 in a magnetic field whose intensity is B = 17 G. Note the electrical polarities at the ends of the wire, and also the polarities of the magnets. What are the direction and magnitude of the magnetic force vector $\mathbf{F}_s$ exerted on a section **s** of this wire 2.7 mm long? Express the answer to two significant figures.

SOLUTION 5-9

Before we do anything else, let's convert all the units to the correct form. We are given these values:

$$I_s = 85 \text{ mA} = 0.085 \text{ A}$$
$$s = 2.7 \text{ mm} = 2.7 \times 10^{-3} \text{ m}$$
$$B = 17 \text{ G} = 1.7 \times 10^{-3} \text{ T}$$

Figure 5-17 shows an angle of 37° between the wire and the magnetic lines of flux. However, we should note that dir **B** is from right to left, while the conventional current, and hence dir **s**, is from lower left to upper right (plus to minus).

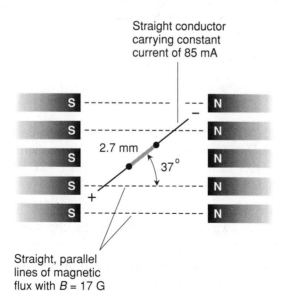

Straight conductor
carrying constant
current of 85 mA

2.7 mm

37°

Straight, parallel
lines of magnetic
flux with B = 17 G

Figure 5-17   Illustration for Problem and
Solution 5-9, and for Quiz Question 6.

The angle θ is meant to be expressed in the rotational sense from **s** toward **B**, so θ is the *supplement* of 37°. That's 180° − 37°, or 143°. The equation for the force on a length of wire tells us this:

$$\mathbf{F_s} = I_s\,(\mathbf{s} \times \mathbf{B})$$

Therefore, dir $\mathbf{F_s}$ is the same as dir $(\mathbf{s} \times \mathbf{B})$. Using the right-hand rule for cross products, we see that this is straight out of the page toward us. Now we're ready to calculate the magnitude of the vector:

$$\begin{aligned} F_s &= I_s\,sB \sin \theta \\ &= 0.085 \times 2.7 \times 10^{-3} \times 1.7 \times 10^{-3} \times \sin 143° \\ &= 2.3 \times 10^{-7}\ \text{N} \end{aligned}$$

Quiz

This is an "open book" quiz. You may refer to the text in this chapter. A good score is 8 correct. Answers are in the back of the book.

1. What is the electrostatic charge quantity carried by a single carbon-atom nucleus that consists of six protons and six neutrons?

    (a) $+9.6 \times 10^{-19}$ C.

    (b) $+1.92 \times 10^{-18}$ C.

    (c) +12 ecu.

    (d) Zero.

2. Observe Fig. 5-16 on page 158 once again. Suppose that the input at the terminals marked "Signal to deflection coils" is an alternating-current (AC) sine wave rather than constant direct current (DC). What will the force vector $\mathbf{F_e}$ be like in that case?

    (a) It will point toward you, just as is described in Solution 5-8.

    (b) It will point away from you, exactly the opposite from the direction described in Solution 5-8.

    (c) It will rotate in a plane perpendicular to the plane of the page, but its magnitude will never change.

    (d) It will point alternately toward and away from you, and will have a magnitude that constantly varies.

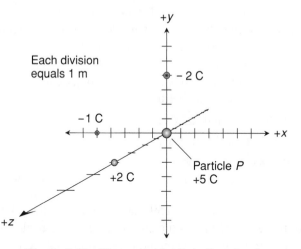

Figure 5-18   Illustration for Quiz Question 3.

3. Imagine a charged particle *P*, surrounded by three other charged particles as shown in the three-space graph of Fig. 5-18. Note that the positive coordinate axes are labeled +*x*, +*y*, and +*z*, and the numbers assigned to the particles represent their charge quantities and polarities. Let **F** be the net force vector imposed on *P* by the other three particles. If we represent **F** as an ordered triple (x,y,z), then based on the fact that like charges repel and opposite charges attract, and also knowing the general concept behind the law of superposition, we can conclude that

(a) $x > 0$, $y > 0$, and $z > 0$.

(b) $x > 0$, $y < 0$, and $z < 0$.

(c) $x < 0$, $y < 0$, and $z < 0$.

(d) $x < 0$, $y > 0$, and $z < 0$.

4. Suppose that the magnetic flux density is 0.64 G at a point 16 cm from a straight wire in free space carrying constant direct current. If the current suddenly doubles, at what distance from the wire will the magnetic flux density be 0.64 G?

(a) 4.0 cm.

(b) 8.0 cm.

(c) 32 cm.

(d) 64 cm.

5. Imagine two tiny particles called P and N that act as point charges. Particle P carries a positive charge, and particle N carries a negative charge. The charge centers are separated by exactly 12 m. Now suppose that the distance between the charge centers is suddenly reduced to 3 m. If the charge on P remains the same, how must the charge on N change so that the electrostatic force is the same as before?

(a) It must become half as great.

(b) It must become 1/4 as great.

(c) It must become 1/16 as great.

(d) We cannot answer this unless we know the initial charge quantities.

6. Observe Fig. 5-17 on page 159 once again. Suppose the magnetic field strength triples to 51 G and the current in the wire triples to 255 mA. Also suppose that the angle between the wire and the magnetic lines of flux does not change. Over what length of wire will the same force magnitude exist as was found in Solution 5-9?

(a) 8.1 mm.

(b) 2.7 mm.

(c) 0.90 mm.

(d) 0.30 mm.

7. Imagine a flat plane X that contains a point charge P. Suppose there are two tiny disk-shaped regions S_1 and S_2 of equal area, both lying entirely in plane X as shown in Fig. 5-19. Let Φ_1 be the electric flux from P that passes through S_1. Let Φ_2 be the electric flux from P that passes through S_2. Suppose that the center of S_1 is d meters from P, and the center of S_2 is kd meters from P, where k is a positive real-number variable. In this scenario, the ratio Φ_1 / Φ_2

(a) is mathematically undefined, because Φ_1 and Φ_2 are both zero.

(b) depends on the value of d.

(c) is equal to k.

(d) is equal to k^2.

8. Suppose the magnetic flux density is a constant 20 mG at a distance of 15 mm from a thin, straight wire in free space that carries direct current. The current in the wire is

(a) 15 mA.

(b) 150 mA.

(c) 6.7 A.

(d) 67 A.

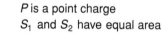

P is a point charge
S_1 and S_2 have equal area

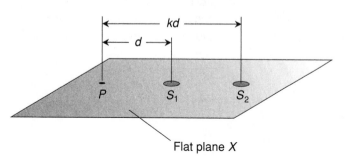

Figure 5-19   Illustration for Quiz Question 7.

9. Consider a small, flat plane region *S* of constant area through which straight lines of electric flux pass at a 90° angle. Suppose the total quantity of flux through *S* is 1.00×10^{-7} C. If *S* is turned so the flux lines pass through at angles of less than 90°, the flux through *S* decreases. When *S* is oriented at a certain angle with respect to the straight lines of flux, the total quantity of flux through *S* is 5.00×10^{-8} C. What is this angle? Assume that the intensity of the electric field in the vicinity of *S* remains constant.

(a) 60°.

(b) 45°.

(c) 30°.

(d) None of the above

10. Consider an evacuated sphere *U* with a point charge *Q* at its center and no other objects inside. Suppose that under these conditions, the total quantity of electric flux passing through *U* is 7.00×10^{-6} C. Suppose the radius of *U* is 100 m, but *U* can be expanded or contracted to any extent at will. What radius will result in a total flux of 7.00×10^{-4} C passing through *U*? Assume that the charge quantity on *Q* remains constant, and that *Q* remains at the center of *U*.

(a) 1.00 cm.

(b) 1.00 m.

(c) 10.0 m.

(d) None of the above

CHAPTER 6

Electrical Impedance and Admittance

Impedance is the extent to which an electrical component or device interferes with the flow of *alternating current* (AC). Its counterpart is *admittance*, the extent to which a component or device allows AC to flow. Before starting this chapter, you should be familiar with electrical *resistance* and *conductance*, and with basic *direct-current* (DC) circuit analysis. These topics are covered in most first-year physics courses, and in Chapter 12 of *Physics Demystified*.

Some Preliminary Math

Before we get into the working details of impedance and admittance, let's review some definitions and principles concerning *imaginary numbers* and *complex numbers*. Engineers use these numbers to model the behavior of electrical components in AC circuits.

IMAGINARY NUMBERS

The value of $(-1)^{1/2}$, the positive square root of -1, is defined as the *unit imaginary number*, symbolized by the lowercase italic j in physics and engineering. The *set of imaginary numbers* is composed of all possible real-number multiples of j. Some examples are $j4$, $j35.79$, $-j25.76$, and $-j25,000$. By convention, the j is written in front of the real number in this format.

The square of an imaginary number is always negative. Some people have trouble grasping this because it is an abstract and unfamiliar concept. But all numbers are abstractions! The imaginary numbers are no less "real" than so-called real numbers such as 4, 35.79, -25.76, or $-25,000$.

The unit imaginary number j can be multiplied by any real number that appears on the *real number line*. If you do this geometric exercise for the whole set of real numbers, you get an *imaginary number line* (Fig. 6-1). The real number line is

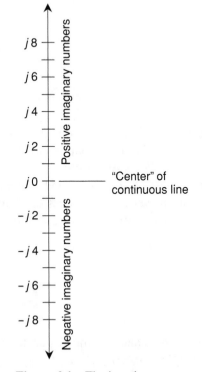

Figure 6-1 The imaginary number line portrays all possible real-number multiples of j.

usually drawn horizontally. The imaginary number line is drawn vertically, per-pendicular to the real-number line.

COMPLEX NUMBERS

When you add a real number and an imaginary number, you get a complex num-ber. In this context, the term "complex" does not mean "complicated." A more descriptive word would be "composite." Examples are $4 + j5$, $8 - j7$, $-7 + j13$, and $-6 - j87$. The set of complex numbers needs a flat two-dimensional *coordinate plane* to be graphically defined.

A complete *complex number plane* is made by taking the real and imaginary number lines and placing them together at right angles so they intersect at the zero points, 0 and $j0$. This is shown in Fig. 6-2. The result is a system of *rectangular coordinates* similar to the grids we use to make graphs such as temperature versus time, or to plot geographic locations according to latitude and longitude.

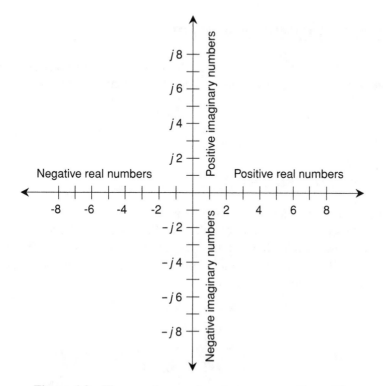

Figure 6-2   The complex number plane portrays all possible sums of real and imaginary numbers.

ADDING AND SUBTRACTING COMPLEX NUMBERS

To add complex numbers, add the real parts and the complex parts separately. For example, the sum of $4 + j7$ and $45 - j83$ works out like this:

$$(4 + j7) + (45 - j83) = (4 + 45) + j(7 - 83)$$
$$= 49 + j(-76)$$
$$= 49 - j76$$

Subtracting complex numbers is a little more involved because the signs can get mixed up. It's best to convert the difference to a sum. For example, the difference $(4 + j7) - (45 - j83)$ can be found by multiplying the second complex number by -1 and then adding the result:

$$(4 + j7) - (45 - j83) = (4 + j7) + [-1 \times (45 - j83)]$$
$$= (4 + j7) + (-45 + j83)$$
$$= (4 - 45) + j(7 + 83)$$
$$= -41 + j90$$

MULTIPLYING COMPLEX NUMBERS

When you multiply complex numbers, treat them as sums of number pairs—that is, as *binomials*. If a, b, c, and d are real numbers, then:

$$(a + jb)(c + jd) = ac + jad + jbc + j^2bd$$
$$= (ac - bd) + j(ad + bc)$$

COMPLEX ORDERED PAIRS

Any complex number can be represented as an *ordered pair* corresponding to a unique point on the complex number plane of Fig. 6-2. In the ordered-pair format for representing a complex number, the real number is written down first, followed by a comma, followed by the imaginary number. Then the whole thing is enclosed in parentheses. Here are some examples:

$$4 + j5 = (4, j5)$$
$$8 - j7 = (8, -j7)$$
$$-7 + j13 = (-7, j13)$$
$$-6 - j87 = (-6, -j87)$$

Unlike the notation for a sequence, list, or set, there should be no space after the comma in the notation for an ordered pair.

COMPLEX NUMBER VECTORS

A complex number can be represented as a vector in the complex number plane. The *originating point*, sometimes called the *back-end point*, of the vector is the coordinate origin $(0, j0)$. The *terminating point*, sometimes called the *end point*, of the vector is the point for the ordered pair (a, jb) representing the number $a + jb$. This gives each complex number vector a unique magnitude and a unique direction.

The magnitude of a complex vector is the distance from the origin to the point (a, jb). The direction of the vector is its angle expressed in the counterclockwise rotational sense from the positive real-number axis. Figure 6-3 shows a generic example. The magnitude of the vector $\mathbf{a + jb}$ representing the complex number $a + jb$ is given by the following formula:

$$|\mathbf{a + jb}| = (a^2 + b^2)^{1/2}$$

When a is positive, the direction of $\mathbf{a + jb}$ is the angle that it subtends with respect to the positive real-number axis, expressed in degrees counterclockwise as follows:

$$\mathrm{dir}\ (\mathbf{a + jb}) = \arctan\ (b/a)$$

When a is negative, the formula is:

$$\mathrm{dir}\ (\mathbf{a + jb}) = 180° + \arctan\ (b/a)$$

If $a = 0$, then dir $(\mathbf{a + jb}) = 90°$ when b is positive, and dir $(\mathbf{a + jb}) = 270°$ when b is negative. If $a = 0$ and $b = 0$, then $(\mathbf{a + jb}) = 0$, and its direction is undefined.

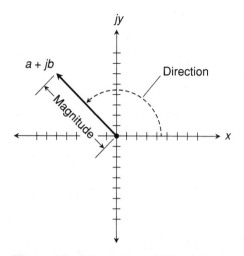

Figure 6-3   Magnitude and direction of a vector in the complex number plane.

ABSOLUTE VALUE

The *absolute value* of a complex number, written $|a + jb|$, is the same as the magnitude of its vector **a + jb** in the complex plane. In the case of a *pure real number* $a + j0$, the absolute value is simply the real number itself, a, if a is positive. If a is negative, then the absolute value of $a + j0$ is equal to $-a$. For a *pure imaginary number* $0 + jb$, the absolute value is equal to b, if b (the real-number *coefficient* of jb) is positive. If b is negative, the absolute value of $0 + jb$ is equal to $-b$. If the number $a + jb$ is neither pure real or pure imaginary, the absolute value must be found by using a formula. First, square both a and b. Then add them. Finally, take the square root. This is the length, c, of the vector **a + jb**. An example is illustrated in Fig. 6-4.

PROBLEM 6-1

Find the absolute value of the complex number $-22 - j0$. Assume the numerical values are exact.

SOLUTION 6-1

This complex number happens to be the same as $-22 + j0$, because $j0 = 0$. Therefore, the absolute value is $-(-22) = 22$.

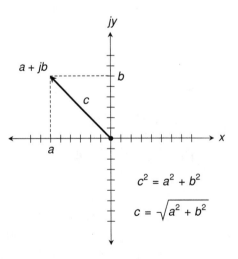

Figure 6-4 Calculation of the absolute value of a complex number in terms of its vector magnitude, represented here by c.

PROBLEM 6-2
Find the absolute value of $0 - j34$. Assume the numerical values are exact.

SOLUTION 6-2
This is a pure imaginary number. The value of b is -34 because $0 - j34 = 0 + j(-34)$. Therefore, the absolute value is $-(-34) = 34$.

PROBLEM 6-3
Find the absolute value of $3 - j4$. Assume the numerical values are exact.

SOLUTION 6-3
Here, $a = 3$ and $b = -4$. We square these numbers, add the results, and then take the square root, as follows:

$$\begin{aligned} |3 - j4| &= [3^2 + (-4)^2]^{1/2} \\ &= (9 + 16)^{1/2} \\ &= 25^{1/2} \\ &= 5 \end{aligned}$$

Impedance

Now that we're armed with imaginary and complex numbers in addition to the old-fashioned "real" ones, we're almost ready to work out a mathematical model that describes how electrical circuits behave with respect to AC. But first, let's review our knowledge of *resistance*, *inductance*, and *capacitance*, and the ways in which these phenomena affect the flow of AC.

RESISTANCE

Resistance, symbolized in equations by the uppercase italic letter R, is a measure of the "brute-force" opposition that a device presents to electric current. The standard unit of resistance is the *ohm*, symbolized by the uppercase Greek letter omega (Ω). You'll also encounter the *kilohm* (kΩ), the *megohm* (MΩ), and the *gigohm* (GΩ), where:

$$1 \text{ k}\Omega = 1000 \ \Omega$$
$$1 \text{ M}\Omega = 1000 \text{ k}\Omega = 10^6 \ \Omega$$
$$1 \text{ G}\Omega = 1000 \text{ M}\Omega = 10^9 \ \Omega$$

When a *potential difference* of one *volt* (1 V) exists across one ohm (1 Ω) of resistance, a current of one *ampere* (1 A) flows through the resistance, assuming the power supply or battery can deliver that much current.

When *n* resistances are connected in series, the total resistance *R* of the combination can be found by adding the values up, as follows:

$$R = R_1 + R_2 + R_3 + \ldots + R_n$$

where R, R_1, R_2, R_3, ..., and R_n are all expressed in the same units. For *n* resistances in parallel, the net resistance *R* of the combination can be found according to the following formula:

$$R = 1 / (1 / R_1 + 1 / R_2 + 1 / R_3 + \ldots + 1 / R_n)$$
$$= (R_1^{-1} + R_2^{-1} + R_3^{-1} + \ldots + R_n^{-1})^{-1}$$

where R, R_1, R_2, R_3, ..., and R_n are all expressed in the same units.

INDUCTANCE

Inductance, symbolized in equations by the uppercase italic letter *L*, is the ability of certain electrical and electronic components to impede the flow of AC by storing energy in the form of a *magnetic field* and releasing that energy later. A device deliberately designed to exhibit electrical inductance is called an *inductor*.

Imagine a resistance-free wire that is extremely long, say 10^6 km. Suppose you fashion this length of wire into a huge loop, and connect its ends to the terminals of a battery. Because the wire is so long, it takes awhile for the electron-motion disturbance, which starts at the negative battery terminal, to make its way around the loop and reach the positive terminal. For this reason, the current starts out low and then increases. Once the disturbance has become uniform around the entire circumference of the loop, the current levels off. The *magnetic flux* produced by the loop is small at first, when current is flowing in only part of the loop. The flux increases as the electron-motion disturbance establishes itself throughout the loop. When the current in the wire finally stabilizes, a certain amount of electrical energy has been stored in the magnetic field produced by the loop. This energy can be later released as an electric current if the battery is disconnected and a *load* such as a resistor or light bulb is put in its place.

Inductors often take the form of wire coils, in effect putting a large wire loop into a small physical space. As the number of turns in a coil increases, so does the inductance, if all other factors are held constant. The material composition of the *core* on which the coil is wound can also affect the inductance by concentrating the magnetic flux. Substances with high magnetic permeability work best for this purpose.

THE HENRY

The unit of inductance is an expression of the ratio between the rate of current change and the instantaneous voltage across an inductor. The standard unit is the *henry*, abbreviated by the uppercase letter H. An inductance of 1 H represents an instantaneous potential difference of 1 V across an inductor within which the instantaneous current is increasing or decreasing at the rate of one *ampere per second* (1 A/s).

The henry is a large unit of inductance. In practice, more common units are the *millihenry* (mH), the *microhenry* (μH), and the *nanohenry* (nH), where:

$$1 \text{ mH} = 0.001 \text{ H} = 10^{-3} \text{ H}$$
$$1 \text{ }\mu\text{H} = 0.001 \text{ mH} = 10^{-6} \text{ H}$$
$$1 \text{ nH} = 0.001 \text{ }\mu\text{H} = 10^{-9} \text{ H}$$

Coils with air cores and few turns of wire have small inductances, in which the current changes quickly and the voltages are low. Coils with ferromagnetic cores and many turns of wire exhibit large inductances, in which the current changes slowly and the voltages are high.

INDUCTANCES IN SERIES AND PARALLEL

Inductances connected in series or parallel add up like resistances in series or parallel, as long as there is no *mutual inductance* among them.

When n inductances are connected in series, the total inductance L of the combination can be found by adding the values up, as follows:

$$L = L_1 + L_2 + L_3 + \ldots + L_n$$

where $L, L_1, L_2, L_3, \ldots$, and L_n are all expressed in the same units.

For n inductances in parallel, the net inductance L of the combination can be found according to the following formula:

$$L = 1 / (1 / L_1 + 1 / L_2 + 1 / L_3 + \ldots + 1 / L_n)$$
$$= (L_1^{-1} + L_2^{-1} + L_3^{-1} + \ldots + L_n^{-1})^{-1}$$

where $L, L_1, L_2, L_3, \ldots$, and L_n are all expressed in the same units.

INDUCTIVE REACTANCE

Inductive reactance is a quantitative expression of the extent that an electrical inductance opposes the flow of AC. It varies with AC frequency, and also with the value of inductance. Inductive reactance is expressed in *imaginary ohms* and is assigned positive values. Denoted jX_L in equations, this quantity can vary from near zero to a few positive-imaginary ohms, to many positive-imaginary kilohms or megohms when the inductance and/or frequency are very high.

If the AC frequency in hertz is f and the value of an inductor in henrys (not "henries"!) is L, then the inductive reactance in imaginary ohms is given by:

$$jX_L = j(2\pi fL)$$

This formula also works for frequencies in megahertz and inductances in microhenrys.

In some texts, inductive reactance is expressed as if it were a real-number quantity, without the unit imaginary number, also called the *j operator*, in front. (Several examples of that notation can be found in *Physics Demystified*, the "little sister" to this book.) This is all right when we talk about imaginary-number inductive reactance alone and don't mix it with real-number resistance. But reactance and resistance often appear together in electrical circuits. We don't want to get them confused. In rigorous usage, therefore, the j operator is included when denoting inductive reactance.

CAPACITANCE

Capacitance, symbolized in equations by the uppercase italic letter C, is the ability of certain electrical and electronic components to impede the flow of AC by storing energy in the form of an *electric field* and releasing that energy later. A device deliberately designed to exhibit electrical capacitance is known as a *capacitor*. In some documentation, such a device is called a *condenser*.

Imagine two flat, identical, gigantic (with surface areas of, say, 10^8 km^2) metal sheets that are both perfect electrical conductors. Suppose they are placed one above the other, separated uniformly by a few centimeters of free space. Now suppose that a battery is connected between these two metal sheets at a corner location. The sheets become charged electrically, one positive and the other negative, but it takes a little while. An electric field arises in the space between the plates. This field is small at first, increases over time, and finally levels off. Electrical energy is stored in this field. The energy can be later released as an electric current if the battery is disconnected and a load is placed between the plates instead.

In general, capacitance is directly proportional to the surface area of the conducting plates or sheets. Capacitance is inversely proportional to the separation between conducting sheets. Capacitance can be increased by stacking or meshing metal plates one atop another, and connecting alternate sets to two electrodes. The capacitance also depends on the nature of the *dielectric* (insulating) material between the plates.

LET THE EXPERIMENTER BEWARE!

In practice, the behavior of a capacitor is similar to the behavior of an inductor. Either type of component can act as an "electrical energy reservoir." The mode of energy storage in a capacitor is vastly different from the mode of energy storage in an inductor. The capacitor stores energy as an E-field, while the inductor stores energy as a B-field. But when that energy is released, the practical effect is the same in either case.

Energy can be stored slowly in a capacitor or inductor, but it can be released in a hurry. Early experimenters in electricity, who built large inductors and capacitors without knowing exactly what they were doing, learned by hard experience that either type of device can store enough energy to deliver a dangerous electric shock if carelessly handled. Never mind whether an E-field or a B-field is the medium of storage. If the energy stored in a large capacitor or inductor happens to be dissipated through your chest cavity all at once, the current surge can kill you!

THE FARAD

In a charging capacitor, the electric field builds up over time from zero to a certain maximum. The greater the capacitance, the slower the rate at which the field builds. The charging process would take longer if the plates were the size of Texas, as compared to the situation if the plates were the size of Delaware. The amount of energy stored in the Texas-sized capacitor would be greater, too.

The unit of capacitance is the *farad*, abbreviated by the uppercase letter F. It expresses the ratio between the instantaneous current and the rate of voltage change between the plates of a charging or discharging capacitor. A capacitance of 1 F represents an instantaneous current flow of 1 A attended by an instantaneous voltage increase or decrease of one *volt per second* (1 V/s).

The farad is a huge unit of capacitance. You'll almost never see a capacitor with a value of 1 F. Commonly employed units of capacitance are the *microfarad* (μF) and the *picofarad* (pF). A capacitance of 1 μF is equal to 10^{-6} F, and 1 pF is equal to 10^{-12} F.

CAPACITANCES IN PARALLEL AND SERIES

Capacitances in parallel combine like resistances or inductances in series, and capacitances in series combine like resistances or inductances in parallel, provided there is no *mutual capacitance* among them.

When n capacitances are connected in parallel, the total capacitance C of the combination can be found by adding the values:

$$C = C_1 + C_2 + C_3 + \ldots + C_n$$

where $C, C_1, C_2, C_3, \ldots$, and C_n are all expressed in the same units.

For n capacitances in series, the net capacitance C of the combination can be found according to this formula:

$$C = 1 / (1 / C_1 + 1 / C_2 + 1 / C_3 + \ldots + 1 / C_n)$$
$$= (C_1^{-1} + C_2^{-1} + C_3^{-1} + \ldots + C_n^{-1})^{-1}$$

where $C, C_1, C_2, C_3, \ldots$, and C_n are all expressed in the same units.

CAPACITIVE REACTANCE

Capacitive reactance is a quantitative expression of the extent to which a capacitor opposes the flow of AC. Capacitive reactance varies with frequency, and also with the value of capacitance. It is expressed in negative-imaginary ohms. Denoted jX_C in equations, capacitive reactance can vary from near zero (when the capacitance and/or the frequency are very high) to a few negative-imaginary ohms, to many negative-imaginary kilohms or megohms when the capacitance and/or frequency are very low.

If the AC frequency in hertz is f and the value of a capacitor in farads is C, then the capacitive reactance in imaginary ohms is given by:

$$jX_C = j / (-2\pi fC) = -j(2\pi fC)^{-1}$$

This formula also works for frequencies in megahertz and capacitances in microfarads.

In some simplified texts, capacitive reactance is expressed without the j operator, as if it were a real-number quantity. This is all right as long as we are dealing with reactance without resistance. But to be rigorous, the j operator should be included when denoting capacitive reactance.

COMPLEX IMPEDANCE

Real-number resistance R and imaginary-number reactance jX (either inductive or capacitive) combine to form *complex impedance*, often symbolized in equations by the italicized uppercase letter Z. Complex impedance is a complete expression of the extent to which a circuit inhibits the flow of AC. Complex impedance values always take the form $R + jX$, where jX is positive for inductive reactance and negative for capacitive reactance. When expressing a complex impedance, the ohm symbol (Ω) is usually left out.

If you're not specifically told the complex impedance (that is, both the resistance and reactance values) when a real-number ohmic figure is quoted as "impedance," it's best to assume that the speaker or author is talking about a *non-reactive impedance*. That means the impedance is a pure resistance. In that case the imaginary, or reactive, quantity is equal to zero.

PROBLEM 6-4

What is the reactance of a 200-pF capacitor at a frequency of 55.4 MHz? Consider the value of π to be 3.14. Express the answer as an imaginary number to three significant figures.

SOLUTION 6-4

First, convert the capacitance value to microfarads: 200 pF = 0.000200 µF. We can now plug in the numbers directly to the formula from above:

$$
\begin{aligned}
jX_C &= -j(2\pi fC)^{-1} \\
&= -j(2.00 \times 3.14 \times 55.4 \times 0.000200)^{-1} \\
&= -j(0.0696^{-1}) \\
&= -j14.4 \ \Omega
\end{aligned}
$$

PROBLEM 6-5

What is the reactance of a 44.00-mH inductor at a frequency of 18.00 kHz? Consider the value of π to be 3.14159. Express the answer in imaginary ohms and imaginary kilohms to four significant figures.

SOLUTION 6-5

First, convert the inductance to microhenrys: 44.00 mH = 4.400×10^4 µH. Then, convert the frequency to megahertz: 18.00 kHz = 0.01800 MHz. We can now plug in the numbers directly to the formula from above:

$$
\begin{aligned}
jX_L &= j(2\pi fL) \\
&= j(2.000 \times 3.14159 \times 0.01800 \times 4.400 \times 10^4) \\
&= j(4.976 \times 10^3) \\
&= j4976 \ \Omega \\
&= j4.976 \ \text{k}\Omega
\end{aligned}
$$

The *RX* Plane

A coordinate system for resistance and inductive reactance is based on the upper-right quadrant of the complex number plane shown in Fig. 6-2. Similarly, a coordinate system for resistance and capacitive reactance is based on the lower-right quadrant of the complex number plane. The two quadrants together form the *resistance-reactance (RX) plane*. The result is actually a half-plane, as shown in Fig. 6-5. Resistance values are represented by non-negative real numbers, and are plotted against the *R* axis. Reactance values, whether inductive (positive) or capacitive (negative), correspond to imaginary numbers, and are plotted against the *jX* axis.

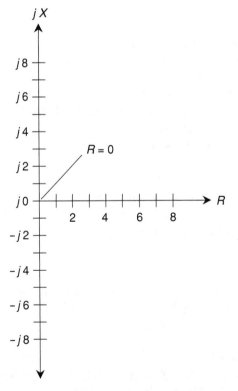

Figure 6-5 The complex impedance plane, also called the resistance-reactance (*RX*) plane.

NO NEGATIVE RESISTANCE

There is no such thing as "negative resistance" in *passive circuits*, that is, electrical networks lacking an external source of energy. You can't have anything better than a perfect conductor. In certain *active circuits*, a DC source such as a battery can be treated as a negative resistance; in other cases, a device can act as if its resistance is negative under changing conditions. But for our purposes here, resistance values are never negative. This is why we can remove the upper-left and lower-left quadrants of the complex number plane and still get a complete set of coordinates for depicting complex impedances in passive circuits.

POINTS AND VECTORS

Think of a point moving around in the *RX* plane, and imagine where the corresponding points on the coordinate axes lie. These points can be found by drawing dashed lines from the point to the *R* and *jX* axes, so that the dashed lines intersect the axes at right angles. Three examples are shown in Fig. 6-6.

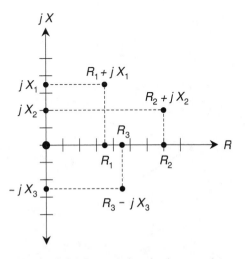

Figure 6-6   Some points in the complex impedance plane, and their resistive and reactive components on the axes.

Now think of points moving toward the right and left, or up and down, on their axes. Imagine what happens to the complex impedance $R + jX$ in various scenarios. This is how the complex impedance changes as the resistance and reactance vary in an electrical circuit. These two parameters can change together, but often they change independently.

Any complex impedance $R + jX$, represented by the ordered pair (R,jX) on the RX plane, can be represented as a vector with its originating point at the coordinate origin and its terminating point at (R,jX). In Fig. 6-7, three complex-impedance vectors are shown on the RX plane, representing the points illustrated in Fig. 6-6. Vectors pointing generally "northeast," or upwards and to the right, represent resistances and inductances connected in series. Vectors pointing in a more or less "southeasterly" direction, or downwards and to the right, represent resistances and capacitances connected in series.

NON-REACTIVE IMPEDANCES

You'll occasionally read or hear that the impedance of some device or component is a certain number of "ohms." For example, in audio electronics, there are "8-Ω" speakers and "600-Ω" amplifier inputs. Figures like these refer to devices that have

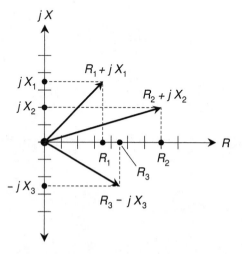

Figure 6-7 Vectors representing the points shown in Fig. 6-6. Illustration for Problem and Solution 6-6.

non-reactive impedance, also known as *purely resistive impedance*. That means an "8-Ω" speaker really has a complex impedance of 8 + j0, and the "600-Ω" input circuit is designed to operate with a complex impedance at, or near, 600 + j0.

Sometimes, the uppercase, italic letter Z is used in place of the word "impedance" in general discussions. In this context, if no specific complex impedance is given, the expression "Z = 8 Ω" can theoretically refer to 8 + j0, 0 + j8, 0 − j8, or any other complex impedance whose point on the *RX* plane lies 8 units from the origin. In practice, such an expression assumes that there is no reactance, so "Z = 8 Ω" would refer to a complex impedance of 8 + j0.

ABSOLUTE-VALUE IMPEDANCE

Suppose an electrical component has a complex impedance of $R + jX$. The *absolute-value impedance*, denoted $|R + jX|$, of the component can be found with the following formula:

$$|R + jX| = (R^2 + X^2)^{1/2}$$

This is sometimes called *simple impedance* and denoted Z. It is the length of the vector in the *RX* plane representing the impedance $R + jX$. But we had better be careful here. We can talk about the length or magnitude of an impedance vector (the absolute value of a complex impedance), saying it is a certain number of "ohms." But there are infinitely many different vectors of any given non-zero length in the *RX* plane. This is why term "ohms" is usually reserved for pure resistance or pure reactance.

PROBLEM 6-6
Examine Fig. 6-7. Suppose that each horizontal division represents an increment of ± 10 Ω, and each vertical division represents an increment of ± j10 Ω. What are the approximate absolute-value impedances represented by the vectors as shown? Make "educated-guess" interpolations for the positions of the component points on the axes. Express the answers to two significant figures.

SOLUTION 6-6
Using our powers of visual interpolation, let's assign these values to the complex impedances shown:

$$R_1 + jX_1 = 35 + j35$$
$$R_2 + jX_2 = 70 + j20$$
$$R_3 - jX_3 = 45 - j25$$

For the first vector we have:

$$
\begin{aligned}
|R_1 + jX_1| &= (R_1^2 + X_1^2)^{1/2} \\
&= (35^2 + 35^2)^{1/2} \\
&= (1225 + 1225)^{1/2} \\
&= 2450^{1/2} \\
&= 49
\end{aligned}
$$

For the second vector we have:

$$
\begin{aligned}
|R_2 + jX_2| &= (R_2^2 + X_2^2)^{1/2} \\
&= (70^2 + 20^2)^{1/2} \\
&= (4900 + 400)^{1/2} \\
&= 5300^{1/2} \\
&= 73
\end{aligned}
$$

For the third vector we have:

$$
\begin{aligned}
|R_3 + jX_3| &= (R_3^2 + X_3^2)^{1/2} \\
&= [45^2 + (-25)^2]^{1/2} \\
&= (2025 + 625)^{1/2} \\
&= 2650^{1/2} \\
&= 51
\end{aligned}
$$

Characteristic Impedance

You'll sometimes hear about *characteristic impedance* or *surge impedance*, symbolized Z_o. This is an important property of *electrical transmission lines*. It is a scalar quantity that can be expressed as a positive real number in ohms.

COMMON TRANSMISSION LINES

Transmission lines at AC frequencies below about 1 GHz usually take either of two forms. Figure 6-8 shows cross-sectional renditions of *coaxial line* (A) and *parallel-wire line*, also called *two-wire line* (B). Examples of transmission lines include the "ribbon" from an outdoor antenna to a hi-fi tuner, the cable from a modem to a computer, and a set of wires that carries high-voltage electricity.

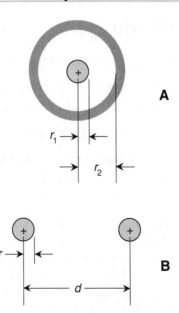

Figure 6-8 Cross-sectional views of coaxial transmission line (A) and parallel-wire line (B). In either type of line, Z_o depends on the conductor diameters and spacing. See text for discussion of the parameters shown.

VOLTAGE, CURRENT, AND Z_o

The characteristic impedance of a transmission line represents the voltage E between the line conductors divided by the current I in the conductors when the line is terminated with an *ideal load*. An ideal load is a pure resistance of just the right ohmic value to allow the most efficient possible transfer of energy from the line to the load. The Z_o also represents the ratio of the voltage to the current in a line of theoretically infinite length, fed with AC at one end. Mathematically:

$$Z_o = E / I$$

where Z_o is in ohms, E is in volts, and I is in amperes.

LINE CONSTRUCTION AND Z_o

If all other factors remain constant in a coaxial transmission line, then:

- The Z_o increases as the center-conductor radius decreases
- The Z_o increases as the inner radius of the outer conductor increases

If all other factors remain constant in a parallel-wire transmission line, then:

- The Z_o increases as the conductor radius decreases
- The Z_o increases as the center-to-center spacing between conductors increases

In Fig. 6-8A for coaxial cable, r_1 is the radius of the center conductor and r_2 is the inner radius of the outer conductor. In drawing B for parallel-wire line, r is the radius of either conductor (it is assumed that the conductors are identical) and d is the center-to-center spacing between the conductors. Given these dimensions, and assuming that the only material between the conductors is free space (dry air or a vacuum), the Z_o of a coaxial cable can be determined approximately from this formula:

$$Z_o = 138 \log (r_2 / r_1)$$

where r_1 and r_2 are expressed in the same units, and "log" represents the base-10 logarithm function. For parallel-wire line, the Z_o can be found as follows:

$$Z_o = 276 \log (d / r)$$

where d and r are expressed in the same units. For any transmission line, the characteristic impedance depends only on its construction. In theory, the Z_o of a coaxial or parallel-wire line is independent of the frequency, and is also independent of the length of the line.

Solid dielectric materials such as *polyethylene* reduce the Z_o of a transmission line, compared with air or a vacuum between the conductors. Because most commercially prefabricated transmission lines employ a solid dielectric between conductors, the Z_o values for typical transmission lines are lower than these formulas imply.

IMPEDANCE MATCHING

In practical applications, the best characteristic impedance for a transmission line in a given situation must be determined according to the nature of the load to which the line delivers power. The optimum or ideal load has a non-reactive impedance equal to the characteristic impedance of the line to which it is connected. If the load has an impedance of $R + j0$, the best possible Z_o for that line is equal to R. In that situation, the current and voltage values along the line exist in the same ratio along the entire length of the line, and the least possible amount of power is dissipated as heat in the line conductors and dielectric.

If the load impedance departs from the Z_o of the transmission line because of a change in the load resistance, the introduction of reactance into the load, or a change in the Z_o of the line itself, the current and voltage values along the line become non-uniform. This causes power to be unnecessarily wasted in the line. *Impedance matching* is the process of making sure that the impedance of a load is purely resistive, with an ohmic value equal to the characteristic impedance of the transmission line connected to it. When the impedances of the line and the load are properly matched, the whole system operates at its peak efficiency.

PROBLEM 6-7

Suppose an infinitely long, two-wire transmission line has a source of 100 V AC connected to one end. The center-to-center spacing between the wires is 2.40 cm, and both wires have a radius of 1.50 mm. The dielectric between the conductors consists entirely of pure, dry air. What is the Z_o of this line? What is the current in the conductors? Now suppose that the center-to-center spacing between the wires is doubled to 4.80 cm without changing the diameter of either wire, and without changing the nature of the dielectric. What is the new Z_o of the line? If the source voltage does not change, what is the new current in either conductor? Express the answers to three significant figures.

SOLUTION 6-7

Before we start calculating, we must be sure the dimensions are all in the same units. We're given the center-to-center spacing d in centimeters but the wire radius r in millimeters. Let's convert d to millimeters so we have $d = 24.0$ and $r = 1.50$. We can plug in the numbers to the formula above to find the characteristic impedance Z_o of this parallel-wire line:

$$
\begin{aligned}
Z_o &= 276 \log (d / r) \\
&= 276 \times \log (24.0 / 1.50) \\
&= 276 \times \log 16.0 \\
&= 332 \ \Omega
\end{aligned}
$$

Because the transmission line is infinitely long, the current in the conductors can be found by rearranging the formula for characteristic impedance in terms of current and voltage. (This can also be done if the line is terminated in a load having a purely resistive impedance of 332 Ω.) We know that $E = 100$ and $Z_0 = 332$. Therefore:

$$Z_0 = E / I$$
$$I = E / Z_0$$
$$= 100 / 332$$
$$= 0.301 \text{ A}$$

We can also call this 301 milliamperes (mA), where 1 mA = 0.001 A.

If the center-to-center spacing is doubled to 4.80 cm or 48.0 mm, then the characteristic impedance increases, as follows:

$$Z_0 = 276 \log (d / r)$$
$$= 276 \times \log (48.0 / 1.50)$$
$$= 276 \times \log 32.0$$
$$= 415 \ \Omega$$

The current will be lower than it was before, assuming the line is infinitely long or is terminated in a purely resistive impedance of 415 Ω, as follows:

$$Z_0 = E / I$$
$$I = E / Z_0$$
$$= 100 / 415$$
$$= 0.241 \text{ A}$$
$$= 241 \text{ mA}$$

Admittance

While impedance is an expression of the extent to which a component, device, or system *discourages* the flow of AC, admittance is a quantification of the extent to which AC is *encouraged* to flow. The relationship between impedance and admittance is the AC analog of the relationship between DC resistance and conductance. But with AC, the mathematics is more involved because we must work in two dimensions rather than in one dimension.

CONDUCTANCE

In AC circuits, electrical conductance plays the same role as it does in DC circuits. In equations, conductance is symbolized by the italic uppercase letter G. The relationship between conductance G and resistance R is simple:

$$G = 1 / R$$

The standard unit of conductance is the *siemens*, abbreviated by the uppercase non-italic letter S. The above equation applies when G is in siemens and R is in ohms.

Conductance is sometimes expressed in *millisiemens* (mS), *microsiemens* (μS), or *nanosiemens* (nS), where:

$$1 \text{ mS} = 0.001 \text{ S}$$
$$1 \text{ μS} = 0.001 \text{ mS} = 10^{-6} \text{ S}$$
$$1 \text{ nS} = 0.001 \text{ μS} = 10^{-9} \text{ S}$$

It is tempting to say that $1 \text{ S} = 1 \text{ Ω}$, $1 \text{ mS} = \text{kΩ}$, $1 \text{ μS} = 1 \text{ MΩ}$, and $1 \text{ nS} = 1 \text{ GΩ}$. For individual components having these values, that's true. But this reasoning can lead to trouble if carelessly applied to circuits with multiple components, because conductances in parallel combine like resistances in series, and conductances in series combine like resistances in parallel.

When n conductances are connected in parallel, the total conductance G of the combination can be found by adding the values up, as follows:

$$G = G_1 + G_2 + G_3 + \ldots + G_n$$

where G, G_1, G_2, G_3, $\ldots$, and G_n are all expressed in the same units. For n conductances in series, the net conductance G of the combination can be found according to the following formula:

$$G = 1 / (1 / G_1 + 1 / G_2 + 1 / G_3 + \ldots + 1 / G_n)$$
$$= (G_1^{-1} + G_2^{-1} + G_3^{-1} + \ldots + G_n^{-1})^{-1}$$

where, again, all the values are expressed in the same units.

SUSCEPTANCE

In AC circuits, *susceptance* is the reciprocal of reactance. It is sometimes abbreviated by the uppercase italic letter B. Don't confuse this with magnetic flux density, which is also customarily symbolized B.

Susceptance is theoretically imaginary like reactance, and can be capacitive (symbolized jB_C) or inductive (symbolized jB_L). Mathematically we have these two relations:

$$jB_C = 1 / jX_C$$
$$jB_L = 1 / jX_L$$

Determining the reciprocals of imaginary numbers is tricky because the reciprocal of j is equal to $-j$. No real number behaves like that! We must keep in mind these two facts:

$$1 / j = -j$$
$$1 / (-j) = j$$

As a result of these properties of j, the sign reverses when we calculate a susceptance value in terms of a reactance value or vice-versa. When expressed in terms of j, inductive susceptance is negative-imaginary, and capacitive susceptance is positive-imaginary—just the opposite situation from inductive reactance and capacitive reactance. That gives us the following equations for inductive and capacitive susceptance in terms of frequency, inductance, and capacitance:

$$jB_L = j / (-2\pi fL) = -j(2\pi fL)^{-1}$$
$$jB_C = j(2\pi fC)$$

where jB_L and jB_C are in imaginary siemens, L is in henrys, and C is in farads. The first formula above will also work if f is in megahertz and L is in microhenrys. The second formula will also work if f is in megahertz and C is in microfarads.

In some simplified texts, susceptance is expressed without the j operator, as if it were a real-number quantity. This is all right as long as we are dealing with susceptance alone, and are not mixing it with real-number conductance. But if we want to be rigorous, the j operator should be included when denoting inductive or capacitive susceptance.

REACTANCE-TO-SUSCEPTANCE CONVERSIONS

Suppose we have an inductive reactance of $j2.45\ \Omega$. To find the inductive susceptance, we must find $1 / (j2.45)$, where the value of 1 is exact. Mathematically, this expression can be converted to a real-number multiple of j in the following manner:

$$1 / (j2.45) = (1 / j)(1.00 / 2.45)$$
$$= (1 / j) \times 0.408$$
$$= -j0.408\ \text{S}$$
$$= -j408\ \text{mS}$$

What about a capacitive reactance of $-j105.7\ \Omega$? To find the capacitive susceptance, we find $1 / (-j105.7)$. Here's how this can be converted to the straightforward product of j and a real number:

$$1 / (-j105.7) = (1 / -j)(1.000 / 105.7)$$
$$= (1 / -j) \times 0.009461$$
$$= j0.009461\ \text{S}$$
$$= j9.461\ \text{mS}$$

COMPLEX ADMITTANCE

Real-number conductance and imaginary-number susceptance combine to form *complex admittance*, symbolized in equations by the italicized uppercase letter *Y*. This is a complete expression of the extent to which a circuit allows AC to flow. In expressions of complex admittance, the siemens symbol (S) is usually left out.

As the absolute value of complex impedance gets larger, the absolute value of complex admittance becomes smaller, in general.

Admittances are written in complex form like impedances. But when you simply write down complex numbers, you need to keep track of which phenomenon you're expressing! This will be obvious if you use the symbol, such as $Y = 3 - j0.5$ or $Y = 7 + j3$. When you see *Y*, you know that negative *j* factors (such as in $3 - j0.5$) mean there is a net inductance in the circuit, and positive *j* factors (such as in $7 + j3$) mean there is net capacitance. When you see *Z*, you know that negative *j* factors (such as in $7 - j50$) mean there is a net capacitance, and positive *j* factors (such as in $78 + j365$) mean there is net inductance.

Admittance is the complex sum, or composite, of conductance and susceptance. Thus, complex admittance values always take the form $Y = G + jB$. When the *j* factor is negative, a complex admittance appears in the form $Y = G - jB$.

PROBLEM 6-8
What is the susceptance of a 200-pF capacitor at 2.540 GHz? Consider the value of π to be 3.14. Express the answer as an imaginary number to three significant figures.

SOLUTION 6-8
Before we do any calculations, we must be sure we have the input data expressed in compatible units. Let's use microfarads and megahertz. Then $C = 0.000200$ and $f = 2540$. Plugging these numbers into the formula for capacitive susceptance, we get:

$$jB_C = j(2\pi f C)$$
$$= j(2.000 \times 3.14 \times 2540 \times 0.000200)$$
$$= j3.19 \text{ S}$$

PROBLEM 6-9
What is the susceptance of a 175.4 nH inductor at a frequency of 144.0 MHz? Consider the value of π to be 3.14159. Express the answer as an imaginary number to four significant figures.

SOLUTION 6-9

Again, we must convert the input data to compatible units before starting with the calculation. Let's use microhenrys and megahertz. That gives us $L = 0.1754$ µH and $f = 144.0$. We plug these values into the formula for inductive susceptance, obtaining:

$$\begin{aligned}
jB_L &= -j(2\pi fL)^{-1} \\
&= -j(2.000 \times 3.14159 \times 144.0 \times 0.1754)^{-1} \\
&= -j(158.7^{-1}) \\
&= -j0.006301 \\
&= -j(6.301 \times 10^{-3}) \text{ S} \\
&= -j6.301 \text{ mS}
\end{aligned}$$

The *GB* plane

Admittance values can be portrayed on a rectangular coordinate plane similar to the complex impedance (*RX*) plane. Actually, it's a half plane, because there is ordinarily no such thing as negative conductance. (You can't have a component that conducts worse than not at all!) Conductance values are plotted against the *G* axis on this coordinate half plane, and susceptance values are plotted against the *jB* axis. Figure 6-9 shows an example of the *GB plane* with several points plotted.

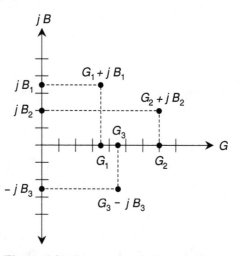

Figure 6-9 Some points in the complex admittance plane, and their conductive and susceptive components on the axes.

In most graphs of this sort, the G axis is oriented horizontally and the jB axis is oriented vertically.

IT'S INSIDE-OUT

To the casual observer, the GB plane looks like the RX plane. But the GB plane is mathematically "inside-out" with respect to the RX plane. The center, or origin, of the GB plane represents the point at which there is no conductance and no susceptance. It is the *zero-admittance point*, rather than the *zero-impedance point*. In the RX plane, the origin represents a *theoretically perfect short circuit*. In the GB plane, the origin represents a *theoretically perfect open circuit*.

As you move out from the origin towards the right ("east") along the G axis of the GB plane, the conductance improves, and the current gets greater. When you move upwards ("north") along the jB axis from the origin, you have ever-increasing positive (capacitive) susceptance. When you go down ("south") along the jB axis from the origin, you encounter increasingly negative (inductive) susceptance.

VECTOR REPRESENTATION OF ADMITTANCE

Complex admittances can be shown as vectors, just as can complex impedances. In Fig. 6-10, the points from Fig. 6-9 are rendered as vectors. As a vector in the GB

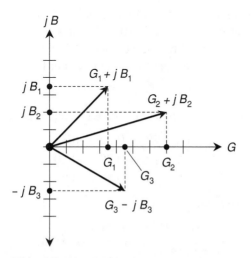

Figure 6-10 Vectors representing the points shown in Fig. 6-9. Illustration for Problem and Solution 6-10.

plane gets longer, the current generally increases; as a vector gets shorter, the current generally decreases. The nature of the *GB* plane gives us a mathematical tool for portraying AC electrical circuits that contain components connected in parallel.

In the *GB* plane, a vector pointing generally "northeast," or upwards and to the right, represents a conductance and a capacitive susceptance connected in parallel. A vector pointing in a more or less "southeasterly" direction, or downwards and to the right, represents a conductance and an inductive susceptance connected in parallel. This is in contrast to vectors in the *RX* plane that represent resistances and capacitive reactances, or resistances and inductive reactances, connected in series.

ABSOLUTE-VALUE ADMITTANCE

Suppose an electrical component has a complex admittance of $G + jB$. The *absolute-value admittance*, denoted $|G + jB|$, of the component can be found with the following formula:

$$|G + jB| = (G^2 + B^2)^{1/2}$$

This is sometimes called *simple admittance* and denoted Y. But here, as with impedance, we must use caution. We can talk about the length or magnitude of an admittance vector, saying it is a certain number of "siemens." But that is ambiguous because there are infinitely many different vectors of any given non-zero length in the *GB* plane.

PROBLEM 6-10

Examine Fig. 6-10. Suppose that each horizontal division represents ± 0.10 S, and each vertical division represents ± j0.10 S. What are the approximate absolute-value admittances represented by the vectors as shown? Make "educated-guess" interpolations. Express the answers to two significant figures.

SOLUTION 6-10

Let's assign the following values by "educated visual guesswork":

$$G_1 + jB_1 = 0.35 + j0.35$$
$$G_2 + jB_2 = 0.70 + j0.20$$
$$G_3 - jB_3 = 0.45 - j0.25$$

For the first vector we have:

$$
\begin{aligned}
|G_1 + jB_1| &= (G_1^2 + B_1^2)^{1/2} \\
&= (0.35^2 + 0.35^2)^{1/2} \\
&= (0.1225 + 0.1225)^{1/2} \\
&= 0.2450^{1/2} \\
&= 0.49
\end{aligned}
$$

For the second vector we have:

$$|G_2 + jB_2| = (G_2{}^2 + B_2{}^2)^{1/2}$$
$$= (0.70^2 + 0.20^2)^{1/2}$$
$$= (0.49 + 0.04)^{1/2}$$
$$= 0.53^{1/2}$$
$$= 0.73$$

For the third vector we have:

$$|G_3 + jB_3| = (G_3{}^2 + B_3{}^2)^{1/2}$$
$$= [0.45^2 + (-0.25)^2]^{1/2}$$
$$= (0.2025 + 0.0625)^{1/2}$$
$$= 0.265^{1/2}$$
$$= 0.51$$

Quiz

This is an "open book" quiz. You may refer to the text in this chapter. A good score is 8 correct. Answers are in the back of the book.

1. Which of the following has a value of $(-16)^{1/2} + 16^{1/2}$? Assume the values given are exact.
   (a) $8 + j0$.
   (b) $4 + j4$.
   (c) $-4 + j4$.
   (d) $0 + j8$.

2. What is the sum of $3 + j2$ and $-3 - j2$? Assume the values given are exact.
   (a) $0 + j0$.
   (b) $6 + j4$.
   (c) $-6 - j4$.
   (d) $0 - j4$.

3. What is the product $(-4.0 - j7.0)(6.0 - j2.0)$ to two significant figures?
   (a) $24 - j14$.
   (b) $-38 - j34$.
   (c) $-24 - j14$.
   (d) $-24 + j14$.

4. The complex admittance value $0.0000 + j0.0355$ represents
   - (a) an essentially pure capacitance.
   - (b) an essentially pure inductance.
   - (c) a conductance, with essentially no capacitance or inductance.
   - (d) a conductance connected in parallel with a capacitance.

5. What is the absolute-value impedance of $3.0 - j6.0$ to two significant figures?
   - (a) 9.0.
   - (b) 3.0.
   - (c) 4.5.
   - (d) 6.7.

6. If the center conductor of a coaxial cable is made to have smaller diameter, all other things being equal, what will happen to the Z_o of the transmission line?
   - (a) It will increase.
   - (b) It will decrease.
   - (c) It will not change.
   - (d) There is no way to determine this without knowing the actual dimensions.

7. Suppose a capacitor has a value of $0.050\ \mu F$ at 665 kHz. What is the capacitive susceptance, stated as an imaginary number to three significant figures?
   - (a) $j4.79$.
   - (b) $-j4.79$.
   - (c) $j0.209$.
   - (d) $-j0.209$.

8. Susceptance and conductance add to form
   - (a) complex impedance.
   - (b) complex inductance.
   - (c) complex reactance.
   - (d) complex admittance.

9. Inductive susceptance is defined in

    (a) imaginary ohms.

    (b) imaginary henrys.

    (c) imaginary farads.

    (d) imaginary siemens.

10. Which of the following statements is false?

    (a) $B_C = 1/X_C$.

    (b) Complex impedance can be depicted as a vector.

    (c) Characteristic impedance is a complex quantity.

    (d) $G = 1/R$.

CHAPTER 7

Alternating-Current Circuit Analysis

We can analyze any network consisting of resistors, inductors, and capacitors with the help of complex numbers and the *RX* (resistance-reactance) or *GB* (conductance-admittance) plane. The *RX* plane applies to series circuits. The *GB* plane applies to parallel circuits. In this chapter, we'll look more closely at the behavior of both types of AC circuits.

Complex Impedances in Series

When you see resistors, inductors, and capacitors connected in series, each component has a complex impedance that can be represented as a vector in the *RX* plane. A vector representing a pure resistance is independent of the frequency. A vector representing a reactance-containing impedance changes if the frequency rises or falls.

PURE REACTANCES

Pure reactances add together in a straightforward fashion when inductors and capacitors are connected in series. That is:

$$jX = jX_L + jX_C$$

where jX is the net reactance, jX_L is the inductive reactance, and jX_C is the capacitive reactance, all expressed in imaginary ohms. In the RX plane, their vectors add. Because these vectors point in exactly opposite directions—inductive reactance straight up and capacitive reactance straight down (Fig. 7-1)—the resultant sum vector points either straight up or straight down unless the reactances are equal and opposite, in which case their sum is the zero vector.

PROBLEM 7-1

Suppose an inductor and capacitor are connected in series, and at the frequency of interest we have $jX_L = j200\ \Omega$ and $jX_C = -j150\ \Omega$. What is the net reactance of this series combination? Express the answer to three significant figures.

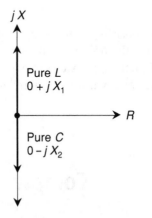

Figure 7-1   Pure inductance and pure capacitance can be represented on the RX plane by vectors that point straight up and straight down, respectively.

SOLUTION 7-1
Add the imaginary numbers, as follows:

$$jX = j200 + (-j150)$$
$$= j(200 - 150)$$
$$= j50 \ \Omega$$

This is a pure inductive reactance because it is positive imaginary.

PROBLEM 7-2
Suppose an inductor and capacitor are connected in series, and at the frequency of interest we have $jX_L = j30 \ \Omega$ and $jX_C = -j110 \ \Omega$. What is the net reactance of this series combination? Express the answer to two significant figures.

SOLUTION 7-2
Again, add the imaginary numbers:

$$jX = j30 + (-j110)$$
$$= j(30 - 110)$$
$$= -j80 \ \Omega$$

This is a pure capacitive reactance because it is negative imaginary.

PROBLEM 7-3
Suppose an inductor whose value is $L = 5.00 \ \mu H$ and a capacitor whose value is $C = 200 \ pF$ are connected in series, and the frequency of operation is $f = 4.00 \ MHz$. What is the net reactance of this series combination? Consider the value of π to be 3.14. Express the final answer to two significant figures.

SOLUTION 7-3
First, calculate the reactance of the inductor at 4.00 MHz. Proceed as follows:

$$jX_L = j2\pi fL$$
$$= j(2.00 \times 3.14 \times 4.00 \times 5.00)$$
$$= j126 \ \Omega$$

Next, calculate the reactance of the capacitor at 4.00 MHz, remembering that 200 pF is the equivalent of 0.000200 μF. Proceed as follows:

$$jX_C = -j(2\pi fC)^{-1}$$
$$= -j(2.00 \times 3.14 \times 4.00 \times 0.000200)^{-1}$$
$$= -j199 \ \Omega$$

Finally, add the inductive and capacitive reactances to obtain the net reactance:

$$jX = jX_L + jX_C$$
$$= j126 + (-j199)$$
$$= j(126 - 199)$$
$$= -j73 \ \Omega$$

This is a pure capacitive reactance.

PROBLEM 7-4

What is the net reactance of the above mentioned inductor and capacitor combination at a frequency of $f = 10.0$ MHz? Consider the value of π to be 3.14. Express the answer to three significant figures.

SOLUTION 7-4

First, calculate the reactance of the inductor at 10.0 MHz. Proceed as follows:

$$jX_L = j2\pi fL$$
$$= j(2.00 \times 3.14 \times 10.0 \times 5.00)$$
$$= j314 \ \Omega$$

Next, calculate the reactance of the capacitor at 10.00 MHz. Proceed as follows:

$$jX_C = -j(2\pi fC)^{-1}$$
$$= -j(2.00 \times 3.14 \times 10.0 \times 0.000200)^{-1}$$
$$= -j79.6 \ \Omega$$

Finally, add the inductive and capacitive reactances to obtain the net reactance:

$$jX = jX_L + jX_C$$
$$= j314 + (-j79.6)$$
$$= j(314 - 79.6)$$
$$= j234 \ \Omega$$

This is a pure inductive reactance.

ADDING COMPLEX IMPEDANCE VECTORS

Suppose two complex impedance vectors don't lie along the same straight line. Figure 7-2 shows an example of such a situation. One vector, which points generally "northeast" in the plot, represents a series resistance-inductance (*RL*) circuit. The other vector, which points generally "southeast," represents a series resistance-capacitance (*RC*) circuit. If you want to add these vectors to determine the net complex impedance vector, you must use the *parallelogram method of vector addition* as shown in Fig. 7-3. Construct a parallelogram using the two vectors

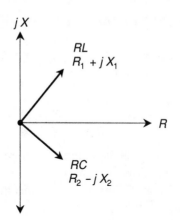

Figure 7-2 When resistance is present along with reactance, the complex impedance vector is neither vertical nor horizontal in the *RX* plane.

$\mathbf{Z}_1 = R_1 + jX_1$ and $\mathbf{Z}_2 = R_2 - jX_2$ as two adjacent sides. The diagonal of the parallelogram is the vector $\mathbf{Z}_1 + \mathbf{Z}_2$ representing the net complex impedance. Note that in a parallelogram, pairs of opposite angles have equal measures. These equalities are indicated by the single and double arcs in Fig. 7-3.

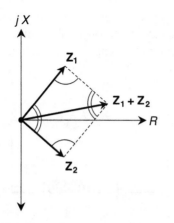

Figure 7-3 Parallelogram method of adding two complex-impedance vectors $\mathbf{Z}_1$ and $\mathbf{Z}_2$.

FORMULA FOR COMPLEX IMPEDANCES IN SERIES

Consider two complex impedances, $\mathbf{Z}_1 = R_1 + jX_1$ and $\mathbf{Z}_2 = R_2 + jX_2$. The net complex impedance vector $\mathbf{Z}$ of these two impedances in series is their vector sum, given by the following formula:

$$\mathbf{Z} = (R_1 + jX_1) + (R_2 + jX_2)$$
$$= (R_1 + R_2) + j(X_1 + X_2)$$

Calculating a vector sum using the formula gives more precise results than doing it geometrically with a parallelogram. The resistance and reactance components add separately. Remember that if a reactance is capacitive, it should be negative-imaginary in this formula.

Series *RLC* Circuits

When a resistance, inductance, and capacitance are connected in series (Fig. 7-4), the resistance R can be imagined as belonging entirely to the inductor, as if it were *ohmic loss* in the inductor wire. Then you have only two complex numbers to add rather than three, when you want to find the net complex impedance vector $\mathbf{Z}$ of the series *resistance-inductance-capacitance* (*RLC*) circuit. In terms of arithmetic addition, we have:

$$\mathbf{Z} = (R + jX_L) + (0 + jX_C)$$
$$= R + j(X_L + X_C)$$

Again, keep in mind that capacitive reactance is always negative imaginary! So, although the formulas here have addition symbols in them, you're adding in a negative number when you "plug in" a capacitive reactance to this formula.

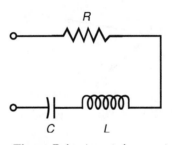

Figure 7-4   A generic
series resistance-inductance-
capacitance (*RLC*) circuit.

PROBLEM 7-5

Suppose a resistor, an inductor, and a capacitor are connected in series with $R = 50\ \Omega$, $jX_L = j22\ \Omega$, and $jX_C = -j33\ \Omega$. What is the net complex impedance? Express the answer with both the real and imaginary components accurate to two significant figures.

SOLUTION 7-5

We can consider the resistor to be part of the inductor. Then we have only two complex impedances to add, $50 + j22$ and $0 - j33$. Adding these yields a resistance component of $50 + 0 = 50\ \Omega$ and a reactance component of $j22 - j33 = -j11\ \Omega$. Therefore, the complex impedance is $50 - j11$.

PROBLEM 7-6

Consider a resistor, an inductor, and a capacitor connected in series with $R = 600\ \Omega$, $jX_L = j444\ \Omega$, and $jX_C = -j444\ \Omega$. What is the net impedance? Express the answer to three significant figures.

SOLUTION 7-6

Again, imagine the resistor to be part of the inductor. Then the complex impedances are $600 + j444$ and $0 - j444$. Adding these, the resistance component is $600 + 0 = 600\ \Omega$, and the reactance component is $j444 - j444 = j0\ \Omega$. The complex impedance is $600 + j0.00$. This is a purely resistive impedance, and we can rightly say "$Z = 600\ \Omega$."

When the inductive and capacitive reactances in a series circuit are both nonzero but they cancel each other out, the circuit is said to be *series-resonant*.

PROBLEM 7-7

Suppose a resistor, inductor, and capacitor are connected in series. The resistance is 330.0 Ω, the capacitance is 220.0 pF, and the inductance is 100.0 µH. The frequency is 7.150 MHz. What is the complex impedance of this series *RLC* circuit at this frequency? Consider the value of π to be 3.14159. Express the answer to four significant figures.

SOLUTION 7-7

First, calculate the inductive reactance. Remember that megahertz and microhenrys go together in the formula. Multiply to obtain the following:

$$jX_L = j2\pi fL$$
$$= j(2.000 \times 3.14159 \times 7.150 \times 100.0)$$
$$= j4492\ \Omega$$

Next, determine the capacitive reactance. Convert 220.0 pF to microfarads, getting $C = 0.0002200\ \mu F$. Then calculate:

$$jX_C = -j(2\pi fC)^{-1}$$
$$= -j(2.000 \times 3.14159 \times 7.150 \times 0.0002200)^{-1}$$
$$= -j101.2\ \Omega$$

Finally, lump the resistance and inductive reactance together, so one of the complex impedances is $330.0 + j4492$. The other is $0.000 - j101.2$. Adding these gives us a net impedance of $330.0 + j4391$.

PROBLEM 7-8
Suppose a resistor, inductor, and capacitor are connected in series. The resistance is $50.0\ \Omega$, the inductance is $10.0\ \mu H$, and the capacitance is 1000 pF. The frequency is 1592 kHz. What is the complex impedance of this series *RLC* circuit at this frequency? Consider the value of π to be 3.14159. Express the answer to three significant figures.

SOLUTION 7-8
First, calculate jX_L, noting that 1592 kHz is equivalent to 1.592 MHz:

$$jX_L = j2\pi fL$$
$$= j(2.00 \times 3.14159 \times 1.592 \times 10.0)$$
$$= j100\ \Omega$$

Next, calculate jX_C. Let's convert picofarads to microfarads, and use megahertz for the frequency. Therefore:

$$jX_C = -j(2\pi fC)^{-1}$$
$$= -j(2.00 \times 3.14159 \times 1.592 \times 0.001000)$$
$$= -j100\ \Omega$$

Put the resistance and inductive reactance together as one vector, $50.0 + j100$. Let the capacitive reactance alone be the other vector, $0 - j100$. The sum of these two vectors is $50.0 + j100 - j100 = 50.0 + j0$, which represents a pure resistance of $50.0\ \Omega$. We can say "$Z = 50.0\ \Omega$" and the circuit is series-resonant at this frequency.

Complex Admittances in Parallel

When you see resistors, inductors, and capacitors in parallel, remember that each component has a complex admittance that can be represented as a vector in the *GB* plane. Any vector representing a pure conductance is independent of the frequency. A vector for an admittance containing susceptance will change if the frequency goes up or down.

PURE SUSCEPTANCES

Pure inductive susceptances (jB_L) and capacitive susceptances (jB_C) add together when inductors and capacitors are connected in parallel. That is:

$$jB = jB_L + jB_C$$

Remember that inductive susceptance is always negative imaginary, and capacitive susceptance is always positive imaginary, the opposite situation from reactances.

In the GB plane, pure jB_L and jB_C vectors add when the respective components are connected in parallel. Because such vectors always point in exactly opposite directions—inductive susceptance straight down and capacitive susceptance straight up—the sum, jB, inevitably points either straight down or straight up (Fig. 7-5), unless the susceptances are equal and opposite in which case they cancel and the result is the zero vector.

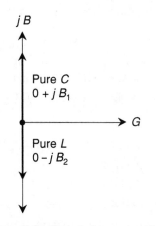

Figure 7-5 Pure inductance and pure capacitance can be represented on the GB plane by vectors that point straight down and straight up, respectively.

PROBLEM 7-9

Suppose an inductor and capacitor are in parallel, with $jB_L = -j0.050$ S and $jB_C = j0.080$ S. What is the net susceptance? Express the answer to two significant figures.

SOLUTION 7-9

Just add the values as follows:

$$jB = jB_L + jB_C$$
$$= -j0.050 + j0.080$$
$$= j0.030 \text{ S}$$

This is a pure capacitive susceptance because it is positive imaginary.

PROBLEM 7-10

Suppose an inductor and capacitor are in parallel, with $jB_L = -j0.60$ S and $jB_C = j0.25$ S. What is the net susceptance? Express the answer to two significant figures.

SOLUTION 7-10

Again, add the values:

$$jB = jB_L + jB_C$$
$$= -j0.60 + j0.25$$
$$= -j0.35 \text{ S}$$

This is a pure inductive susceptance because it is negative imaginary.

PROBLEM 7-11

Suppose an inductor of $L = 6.00$ μH and a capacitor of $C = 150$ pF are connected in parallel. The frequency is $f = 4.00$ MHz. What is the net susceptance? Consider the value of π to be 3.14. Express the answer to three significant figures.

SOLUTION 7-11

First, calculate the susceptance of the inductor at 4.00 MHz, as follows:

$$jB_L = -j(2\pi fL)^{-1}$$
$$= -j(2.00 \times 3.14 \times 4.00 \times 6.00)^{-1}$$
$$= -j0.00663 \text{ S}$$

Next, calculate the susceptance of the capacitor (converting its value to microfarads) at 4.00 MHz, as follows:

$$jB_C = j2\pi fC$$
$$= j(2.00 \times 3.14 \times 4.00 \times 0.000150)$$
$$= j0.00377 \text{ S}$$

Finally, add the inductive and capacitive susceptances to obtain the net susceptance:

$$jB = jB_L + jB_C$$
$$= -j0.00663 + j0.00377$$
$$= -j0.00286 \text{ S}$$

This is a pure inductive susceptance.

PROBLEM 7-12
What is the net susceptance of the above mentioned parallel-connected inductor and capacitor at a frequency of $f = 5.31$ MHz? Consider the value of π to be 3.14. Express the answer to three significant figures.

SOLUTION 7-12
First, calculate the susceptance of the inductor at 5.31 MHz, as follows:

$$jB_L = -j(2\pi fL)^{-1}$$
$$= -j(2.00 \times 3.14 \times 5.31 \times 6.00)^{-1}$$
$$= -j0.00500 \text{ S}$$

Next, calculate the susceptance of the capacitor (converting its value to micro-farads) at 5.31 MHz, as follows:

$$jB_C = j2\pi fC$$
$$= j(2.00 \times 3.14 \times 5.31 \times 0.000150)$$
$$= j0.00500 \text{ S}$$

Finally, add the inductive and capacitive susceptances to obtain the net susceptance:

$$jB = jB_L + jB_C$$
$$= -j0.00500 + j0.00500$$
$$= j0.00 \text{ S}$$

The circuit has zero susceptance at 5.31 MHz. When the inductive and capacitive susceptances in a parallel circuit are both non-zero but they cancel as they do here, the circuit is said to be *parallel-resonant*.

ADDING COMPLEX ADMITTANCE VECTORS

When the conductance is significant in a parallel circuit containing inductance and capacitance, the admittance vectors do not point straight up and down. Instead,

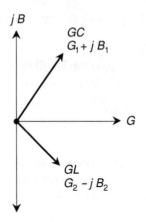

Figure 7-6 When conductance is present along with susceptance, the complex admittance vector is neither vertical nor horizontal in the *GB* plane.

they run off generally "northeast" (for the capacitive part of the circuit) and generally "southeast" (for the inductive part). Figure 7-6 shows a generic example of this. We've seen how vectors add in the *RX* plane. In the *GB* plane, the principle is the same. The net admittance vector is the sum of the component admittance vectors. We can use the parallelogram method of vector addition in the coordinate plane, or add the complex numbers mathematically.

FORMULA FOR COMPLEX ADMITTANCES IN PARALLEL

Given two admittances, $\mathbf{Y}_1 = G_1 + jB_1$ and $\mathbf{Y}_2 = G_2 + jB_2$, the net admittance vector $\mathbf{Y}$ of these in parallel is their vector sum, as follows:

$$\mathbf{Y} = (G_1 + jB_1) + (G_2 + jB_2)$$
$$= (G_1 + G_2) + j(B_1 + B_2)$$

The conductance and susceptance components add separately. Remember again that if a susceptance is inductive, then it is negative imaginary in this formula.

Parallel *GLC* Circuits

When a conductance, inductance, and capacitance are connected in parallel (Fig. 7-7), we can imagine the conductance as existing entirely in the capacitor, as if it were a *leakage conductance* in the dielectric material between the plates. Then we have only two, rather than three, vectors to add when finding the complex admittance vector for a parallel *GLC* (conductance-inductance-capacitance) circuit:

$$\mathbf{Y} = (0 + jB_L) + (G + jB_C)$$
$$= G + j(B_L + B_C)$$

Again, that old reminder: the value of B_L is always negative! So, although the formulas here have addition symbols in them, we're adding a negative number when we add in an inductive susceptance.

PROBLEM 7-13

Suppose a resistor, inductor, and capacitor are connected in parallel. Suppose the resistor has a conductance $G = 0.10$ S, and the susceptances are $jB_L = -j0.010$ S and $jB_C = j0.020$ S. What is the complex admittance of this combination? Express the answer to two significant figures.

SOLUTION 7-13

Consider the resistor to be part of the capacitor. Then there are two complex admittances in parallel: $0.00 - j0.010$ and $0.10 + j0.020$. Adding these gives a conductance component of $0.00 + 0.10 = 0.10$ S and a susceptance component of $-j0.010 + j0.020 = j0.010$ S. Therefore, the complex admittance is $0.10 + j0.010$.

PROBLEM 7-14

Suppose a resistor, inductor, and capacitor are connected in parallel. The resistor has conductance $G = 0.0010$ S, and the susceptances are $jB_L = -j0.0022$ S and

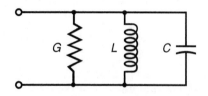

Figure 7-7   A generic parallel conductance-inductance-capacitance (*GLC*) circuit.

$jB_C = j0.0022$ S. What is the complex admittance of this combination? Express the answer to two significant figures.

SOLUTION 7-14

Again, consider the resistor to be part of the capacitor. Then the complex admittances are $0.0000 - j0.0022$ and $0.0010 + j0.0022$. Adding these, the conductance component is $0.0000 + 0.0010 = 0.0010$ S, and the susceptance component is $-j0.0022 + j0.0022 = j0.0$ S. Thus, the admittance is $0.0010 + j0.0$. This is a purely conductive admittance. The circuit is parallel-resonant because the susceptances, both non-zero, cancel.

PROBLEM 7-15

Suppose a resistor, inductor, and capacitor are connected in parallel. The resistance is 100 Ω, the capacitance is 200 pF, and the inductance is 100 μH. The frequency is 1.00 MHz. What is the net complex admittance? Consider the value of π to be 3.14. Express the answer to two significant figures.

SOLUTION 7-15

First, calculate the inductive susceptance. Recall the formula, and plug in the numbers as follows, noting that megahertz go with microhenrys:

$$jB_L = -j(2\pi fL)^{-1}$$
$$= -j(2.00 \times 3.14 \times 1.00 \times 100)^{-1}$$
$$= -j0.00159 \text{ S}$$

Next, calculate the capacitive susceptance. Convert 200 pF to microfarads to go with megahertz in the formula, getting $C = 0.000200$ μF. Then:

$$jB_C = j2\pi fC$$
$$= j(2.00 \times 3.14 \times 1.00 \times 0.000200)$$
$$= j0.00126 \text{ S}$$

Finally, consider the conductance, which is the reciprocal of the resistance. We have:

$$G = 1/100$$
$$= 0.010 \text{ S}$$

Let's think of this conductance and the capacitive susceptance as existing together in a single component. That means that one of the admittances is $0.0000 - j0.00159$, and the other is $0.010 + j0.00126$. Adding these gives a net admittance of $0.010 - j0.00033$ for this circuit at this frequency.

PROBLEM 7-16

Suppose a resistor, inductor, and capacitor are in parallel. The resistance is 10.0 Ω, the inductance is 10.0 µH, and the capacitance is 1000 pF. The frequency is 1592 kHz. What is the complex admittance of this circuit at this frequency? Consider the value of π to be 3.14159. Express the answer to three significant figures.

SOLUTION 7-16

First, calculate the inductive susceptance. Convert the frequency to megahertz, obtaining f = 1.592 MHz. Then plug in the numbers as follows:

$$jB_L = -j(2\pi fL)^{-1}$$
$$= -j(2.00 \times 3.14159 \times 1.592 \times 10.0)^{-1}$$
$$= -j0.0100 \text{ S}$$

Next, calculate the capacitive susceptance. Convert 1000 pF to microfarads to go with megahertz in the formula, getting C = 0.001000 µF. Then:

$$jB_C = j2\pi fC$$
$$= j(2.00 \times 3.14159 \times 1.592 \times 0.001000)$$
$$= j0.0100 \text{ S}$$

Finally, consider the conductance, which is the reciprocal of the resistance. We have:

$$G = 1/10.0$$
$$= 0.100 \text{ S}$$

which we'll consider as part of the capacitor. Thus, one of the parallel-connected admittances is 0.00 − j0.0100, and other is 0.100 + j0.0100. Adding gives 0.100 + j0.00. Because the net susceptance is zero, this circuit is parallel-resonant at 1592 kHz.

CONVERTING ADMITTANCE TO IMPEDANCE

Once we know the admittance of a parallel *GLC* circuit, we can transform it to an impedance. The conversion from a complex admittance $G + jB$ to a complex impedance $R + jX$ requires two formulas: one for R, the real-number resistance, and the other for X, the real-number coefficient of the imaginary-number reactance. Here they are:

$$R = G / (G^2 + B^2)$$
$$X = -B / (G^2 + B^2)$$

If we know the complex admittance, we must find the resistance and reactance components individually using the above formulas. Then we can assemble the two components into the complex impedance, $R + jX$. Let's try an example.

PROBLEM 7-17

Suppose the complex admittance of a certain parallel circuit is $0.010 - j0.0050$. What is the complex impedance of this same circuit, assuming the frequency does not change? Express the answer to two significant figures.

SOLUTION 7-17

In this case, $G = 0.010$ and $B = -0.0050$. First find $G^2 + B^2$, as follows:

$$G^2 + B^2 = 0.010^2 + (-0.0050)^2$$
$$= 0.000100 + 0.000025$$
$$= 0.000125$$

Now it is easy to calculate R and X, like this:

$$R = G / 0.000125$$
$$= 0.010 / 0.000125$$
$$= 80$$

$$X = -B / 0.000125$$
$$= -(-0.0050) / 0.000125$$
$$= 0.0050 / 0.000125$$
$$= 40$$

The complex impedance is therefore $R + jX = 80 + j40$.

Putting It All Together

When you're confronted with a parallel *GLC* circuit and you want to determine the complex impedance vector **Z** of the combination, perform the following operations in this order:

1. Find the real-number conductance $G = 1 / R$ of the resistor.
2. Find the imaginary-number susceptance jB_L of the inductor.
3. Find the imaginary-number susceptance jB_C of the capacitor.

4. Find the net imaginary-number susceptance $jB = jB_L + jB_C$.

5. Determine the real-number value $G^2 + B^2$.

6. Compute the real-number values R and X in terms of the real-number values G and B, paying careful attention to the signs.

7. Assemble the complex impedance $R + jX$.

PROBLEM 7-18

Suppose a resistor of 10.0 Ω, a capacitor of 820 pF, and an inductor of 10.0 µH are in parallel. The frequency is 1.00 MHz. What is the complex impedance? Consider the value of π to be 3.14159. Express the final answer to three significant figures.

SOLUTION 7-18

Let's proceed according to the sequence of steps outlined above, paying careful attention to notation, exponents, and signs. At intermediate steps in the calculation after π is introduced, we will go to six significant digits (because that is the accuracy to which we are told the value of π), rounding to three significant figures at the end. This will prevent cumulative errors that can result from repeated rounding of values.

1. Find the real-number conductance $G = 1 / R$ of the resistor:

$$G = 1 / R$$
$$= 1 / 10.0$$
$$= 0.100 \text{ S}$$

2. Find the imaginary-number susceptance jB_L of the inductor:

$$jB_L = -j(2\pi fL)^{-1}$$
$$= -j(2.00 \times 3.14159 \times 1.00 \times 10.0)^{-1}$$
$$= -j0.0159155 \text{ S}$$

3. Find the imaginary-number susceptance jB_C of the capacitor after converting the capacitance to microfarads to go with megahertz:

$$jB_C = j2\pi fC$$
$$= j(2.00 \times 3.14159 \times 1.00 \times 0.000820)$$
$$= j0.00515221 \text{ S}$$

4. Find the net imaginary susceptance:

$$jB = jB_L + jB_C$$
$$= -j0.0159155 + j0.00515221$$
$$= -j0.0107633 \text{ S}$$

5. Determine the value of $G^2 + B^2$:

$$G^2 + B^2 = 0.100^2 + (-0.0107633)^2$$
$$= 0.010000 + 0.000115849$$
$$= 0.0101158$$

6. Compute the real-number values R and X in terms of the real-number values G and B, paying careful attention to the signs and rounding off each value to three significant figures:

$$R = G / 0.0101158$$
$$= 0.100 / 0.0101158$$
$$= 9.89$$

$$X = -B / 0.0101158$$
$$= -(-0.0107633) / 0.0101158$$
$$= 1.06$$

7. Assemble the complex impedance:

$$R + jX = 9.89 + j1.06$$

PROBLEM 7-19

Suppose a resistor of 47.0 Ω, a capacitor of 500 pF, and an inductor of 10.0 μH are in parallel. What is their complex impedance at a frequency of 2.252 MHz? Consider the value of π to be 3.14159. Express the answer to three significant figures.

SOLUTION 7-19

Again, let's perform the sequence of steps outlined above, paying careful attention to notation, exponents, and signs. Again, we'll allow some extra significant digits before reaching the end of the calculation to prevent cumulative rounding errors.

1. Find the real-number conductance $G = 1 / R$ of the resistor:

$$G = 1 / R$$
$$= 1 / 47.0$$
$$= 0.0212766 \text{ S}$$

2. Find the imaginary-number susceptance jB_L of the inductor:

$$jB_L = -j(2\pi fL)^{-1}$$
$$= -j(2.00 \times 3.14159 \times 2.252 \times 10.0)^{-1}$$
$$= -j0.00706728 \text{ S}$$

3. Find the imaginary-number susceptance jB_C of the capacitor after converting the capacitance to microfarads to go with megahertz:

$$jB_C = j2\pi fC$$
$$= j(2.00 \times 3.14159 \times 2.252 \times 0.000500)$$
$$= j0.00707486 \text{ S}$$

4. Find the net imaginary susceptance:

$$jB = jB_L + jB_C$$
$$= -j0.00706728 + j0.00707486$$
$$= j0.00000758 \text{ S}$$
$$= j(7.58 \times 10^{-6}) \text{ S}$$

5. Determine the value of $G^2 + B^2$:

$$G^2 + B^2 = 0.0212766^2 + (7.58 \times 10^{-6})^2$$
$$= (4.52694 \times 10^{-4}) + (5.74564 \times 10^{-11})$$
$$= 4.52694 \times 10^{-4}$$

For all practical purposes, the value of B^2 vanishes here; it is too small to be of any consequence in the sum $G^2 + B^2$. That's why we ignore it in the result.

6. Compute the real-number values R and X in terms of the real-number values G and B, paying attention to the signs and rounding off each value to three significant figures:

$$R = G / (4.52694 \times 10^{-4})$$
$$= 0.0212766 / (4.52694 \times 10^{-4})$$
$$= 47.0$$

$$X = -B / (4.52694 \times 10^{-4})$$
$$= -(7.58 \times 10^{-6}) / (4.52694 \times 10^{-4})$$
$$= -0.0167$$

7. Assemble the complex impedance:

$$R + jX = 47.0 - j0.0167$$

At 2.252 MHz, this circuit presents an almost perfectly pure resistance equal to the value of the resistor, indicating that the circuit is nearly parallel-resonant.

Ohm's Law for AC Circuits

Ohm's law for DC circuits is a simple relationship among voltage E, current I, and resistance R that can be expressed in three ways:

$$E = IR$$
$$I = E / R$$
$$R = E / I$$

where E is in volts, I is in amperes, and R is in ohms. Things get a little more complicated in AC circuits.

ROOT-MEAN-SQUARE (RMS)

Often, it is necessary to express the *effective amplitude* of an AC wave. This is the voltage or current that a DC source would have to produce in order to "have the same effect" as a given AC wave. When you say that a wall outlet provides 117 V, you mean 117 *effective volts*. The most common expression for effective AC intensity is called the *root-mean-square (rms) amplitude*. The terminology reflects the fact that the AC wave is mathematically "operated on" by taking the square root of the mean (average) of the squares of all its instantaneous amplitudes.

In the case of a perfect sine wave, which is the only type of AC wave we're concerned with here, the rms amplitude is equal to 0.707 times the *peak amplitude*, or 0.354 times the *peak-to-peak amplitude*.

OHM'S LAW FOR COMPLEX IMPEDANCES

When the impedance in an AC circuit is either a pure resistance R or a pure reactance jX, Ohm's law can be expressed as follows:

$$E = IR \text{ or } E = IX$$
$$I = E / R \text{ or } I = E / X$$
$$R = E / I \text{ or } X = E / I$$

where E is the alternating voltage in volts rms, I is the alternating current in amperes rms, R is the resistance in ohms, and X is the real-number coefficient of the imaginary-number reactance jX. Note that R is independent of frequency, but X is not! This is why, when you want to determine the relationship among current, voltage, and resistance in an AC circuit that contains both resistance and reactance, things get a little messy.

In the previous chapter, you learned that the absolute-value impedance Z of a component with resistance R and reactance jX is:

$$Z = |R + jX|$$
$$= (R^2 + X^2)^{1/2}$$

This formula applies for series RLC circuits. The absolute-value impedance Z for parallel RLC circuits can be determined as well. The formula turns out like this:

$$Z = [R^2 X^2 / (R^2 + X^2)]^{1/2}$$

These formulas allow us to generalize Ohm's law for AC circuits so it is valid even if the impedance contains both resistance and reactance. The general formulas look like this:

$$E = IZ$$
$$I = E / Z$$
$$Z = E / I$$

For series circuits, substitution gives us:

$$E = I (R^2 + X^2)^{1/2}$$
$$I = E / [(R^2 + X^2)^{1/2}]$$
$$R^2 + X^2 = E^2 / I^2$$

For parallel circuits, substitution gives us:

$$E = I [R^2 X^2 / (R^2 + X^2)]^{1/2}$$
$$I = E / \{[R^2 X^2 / (R^2 + X^2)]^{1/2}\}$$
$$R^2 X^2 / (R^2 + X^2) = E^2 / I^2$$

where E is the alternating voltage in volts rms, I is the alternating current in amperes rms, R is the resistance in ohms, and X is the real-number coefficient of the reactance jX in imaginary ohms.

PROBLEM 7-20
Suppose a series RX circuit, shown by the generic block diagram of Fig. 7-8, has a resistance of $R = 50.00\ \Omega$ and a capacitive reactance of $jX = -j50.00\ \Omega$ at a certain frequency. Suppose 100.0 V rms AC is applied to this circuit at that frequency. What is the rms current? Express the answer to four significant figures.

SOLUTION 7-20
First, calculate the absolute-value impedance:

$$Z = (R^2 + X^2)^{1/2}$$
$$= [50.00^2 + (-50.00)^2]^{1/2}$$
$$= (2500 + 2500)^{1/2}$$
$$= 5000^{1/2}$$
$$= 70.71$$

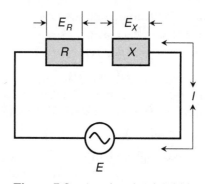

Figure 7-8 A series circuit with "gray boxes" containing resistance and reactance. Illustration for Problems 7-20 through 7-23.

Then calculate the current using Ohm's law for AC circuits. We have:

$$I = E / Z$$
$$= 100.0 / (5000)^{1/2}$$
$$= 100.0 / 70.71$$
$$= 1.414 \text{ A rms}$$

PROBLEM 7-21

What are the rms AC voltages across the resistance and the reactance, respectively, in the circuit described in Problem 7-20? Express the answers to four significant figures.

SOLUTION 7-21

Because the current is $I = 1.414$ A rms, the voltage drop E_R across the resistance is:

$$E_R = IR$$
$$= 1.414 \times 50.00$$
$$= 70.70 \text{ V rms}$$

The voltage drop E_X across the reactance is:

$$E_X = IX$$
$$= 1.414 \times (-50.0)$$
$$= -70.70 \text{ V rms}$$

This is an rms AC voltage of equal magnitude to that across the resistance. But the *phase* is different. The waves aren't "in sync." The sign attached to E_X has meaning only with respect to the reactive part of the circuit; it is irrelevant with respect to the resistive part.

If you're particularly astute, you will notice that the values 70.71, 70.70, and −70.70 that we obtained in these calculations represent $5000^{1/2}$, $5000^{1/2}$, and $−(5000^{1/2})$, respectively. The slight difference in the final figures arises among these examples because of cumulative rounding errors. (Sometimes, it seems, we can't get away from this bugaboo!)

REACTANCE AND PHASE

Note that voltages across the resistance and the reactance—a capacitive reactance in this case, because it's negative—don't add up to 100 V rms, which is placed across the whole circuit. This is no artifact of cumulative rounding errors. It's not just off by a little bit; it's way off!

The reason for this apparent discrepancy is the fact that in an AC circuit containing resistance and reactance, there is always a difference in phase between the voltage across the resistance and the voltage across the reactance. The voltages across the components always add up to the applied voltage *vectorially*, but not *arithmetically*.

We won't delve into the idiosyncrasies of phase vectors in complex AC circuits here, but it bears mention in passing so you aren't left wondering if something is wrong with the whole theory of AC circuits.

PROBLEM 7-22
Suppose a series *RX* circuit of the sort shown in Fig. 7-8 has a resistance of $R = 10.0\ \Omega$ and an inductive reactance of $jX = j40.0\ \Omega$ at a certain frequency. The applied voltage is 100 V rms AC. What is the current? Express the answer to three significant figures.

SOLUTION 7-22
First, calculate the absolute-value impedance:

$$
\begin{aligned}
Z &= (R^2 + X^2)^{1/2} \\
&= [10.0^2 + 40.0^2]^{1/2} \\
&= (100 + 1600)^{1/2} \\
&= 1700^{1/2} \\
&= 41.2
\end{aligned}
$$

Then calculate the current using Ohm's law for AC circuits:

$$I = E / Z$$
$$= 100 / 41.2$$
$$= 2.43 \text{ A rms}$$

PROBLEM 7-23
What are the rms AC voltages across the resistance and the reactance, respectively, in the circuit described in Problem 7-22 and illustrated in Fig. 7-8? Express the answers to three significant figures.

SOLUTION 7-23
Knowing the current, we can calculate the voltage across the resistance:

$$E_R = IR$$
$$= 2.43 \times 10.0$$
$$= 24.3 \text{ V rms}$$

We can calculate the voltage across the reactance as follows:

$$E_X = IX$$
$$= 2.43 \times 40.0$$
$$= 97.2 \text{ V rms}$$

If you add $E_R + E_X$ arithmetically, you don't get the applied voltage! The simple DC rule does not work here for the same reason it doesn't work in Problem 7-21.

PROBLEM 7-24
Suppose the parallel RX circuit shown in Fig. 7-9 has a resistance of $R = 30.0 \ \Omega$ and a reactance $jX = -j20.0 \ \Omega$. The AC supply voltage is 50.0 V rms. What is the total current drawn from the AC supply? Express the answer to three significant figures.

SOLUTION 7-24
First, find the absolute-value impedance, remembering the formula for parallel circuits:

$$Z = [R^2 X^2 / (R^2 + X^2)]^{1/2}$$
$$= \{30.0^2 \times (-20.0)^2 / [30.0^2 + (-20.0)^2]\}^{1/2}$$
$$= [900 \times 400 / (900 + 400)]^{1/2}$$
$$= [(3.60 \times 10^5) / 1300]^{1/2}$$
$$= 277^{1/2}$$
$$= 16.6$$

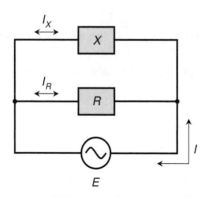

Figure 7-9 A parallel circuit with "gray boxes" containing resistance and reactance. Illustration for Problems 7-24 and 7-25.

The total current is therefore:

$$I = E / Z$$
$$= 50.0 / 16.6$$
$$= 3.01 \text{ A rms}$$

PROBLEM 7-25

What are the rms currents through the resistance and the reactance, respectively, in the circuit described in Problem 7-24? Express the answers to three significant figures.

SOLUTION 7-25

Knowing the voltage, we can calculate the current through the resistance:

$$I_R = E / R$$
$$= 50.0 / 30.0$$
$$= 1.67 \text{ A rms}$$

We can calculate the current through the reactance as follows:

$$I_X = E / X$$
$$= 50.0 / (-20.0)$$
$$= -2.50 \text{ A rms}$$

Note that these currents don't add up to 3.01 A rms, the total current actually drawn from the source. The reason for this is the same as the reason the rms voltages don't add arithmetically in AC circuits that contain reactance. The constituent currents, I_R and I_X, differ in phase. Vectorially, they add up to 3.01 A rms, but arithmetically they don't.

Quiz

This is an "open book" quiz. You may refer to the text in this chapter. A good score is 8 correct. Answers are in the back of the book.

1. Suppose an inductor and capacitor are connected in series. The inductive reactance is $j250\ \Omega$, and the capacitive reactance is $-j300\ \Omega$. What is the complex impedance?

   (a) $0 + j550$.

   (b) $0 - j50$.

   (c) $250 - j300$.

   (d) $-300 + j250$.

2. When $R = 0$ and $jX = j5$ in a series RLC circuit, the complex impedance vector

   (a) points straight up.

   (b) points straight down.

   (c) points straight towards the right.

   (d) None of the above

3. Suppose a resistor of $330\ \Omega$, an inductor of $1.00\ \mu H$, and a capacitor of $200\ pF$ are in series. What is the complex impedance at $10.0\ MHz$?

   (a) $330 - j199$.

   (b) $300 + j201$.

   (c) $300 + j142$.

   (d) $330 - j16.8$.

4. Suppose an inductor has a pure reactance of $j4.00$ Ω. What is the complex admittance?

   (a) $0 + j0.25$.

   (b) $0 + j4.00$.

   (c) $0 - j0.25$.

   (d) $0 - j4.00$.

5. Consider an inductor and capacitor in parallel, with $jB_L = -j0.05$ and $jB_C = j0.03$. What is the complex admittance?

   (a) $0 - j0.02$.

   (b) $0 - j0.07$.

   (c) $0 + j0.02$.

   (d) $-0.05 + j0.03$.

6. Suppose an inductor of 3.50 μH and a capacitor of 47.0 pF are in parallel. The frequency is 9.55 MHz. What is the complex admittance?

   (a) $0 + j0.00282$.

   (b) $0 - j0.00194$.

   (c) $0 + j0.00194$.

   (d) $0 - j0.00758$.

7. Suppose a conductance of 0.0044 S, a capacitive susceptance of $j0.035$ S, and an inductive susceptance of $-j0.011$ S are in parallel. What is the complex admittance?

   (a) $0.0044 + j0.024$.

   (b) $0.035 - j0.011$.

   (c) $-0.011 + j0.035$.

   (d) $0.0044 + j0.046$.

8. Suppose the complex admittance of a parallel GLC circuit is $0.02 + j0.20$. What is the complex impedance, assuming the frequency does not change?

   (a) $50 + j5.0$.

   (b) $0.495 - j4.95$.

   (c) $50 - j5.0$.

   (d) $0.495 + j4.95$.

9. Suppose a series circuit has 99.0 Ω of resistance and $j88.0$ Ω of inductive reactance at a certain frequency f. An AC voltage of 117 V rms having frequency f is applied to this series network. What is the current drawn from the source?

   (a) 1.60 A rms.

   (b) 1.13 A rms.

   (c) 0.883 A rms.

   (d) 0.626 A rms.

10. Suppose a parallel circuit has 10 Ω of resistance and 15 Ω of reactance at a certain frequency f. An AC voltage of 20 V rms having frequency f is applied across this network. What is the current drawn from the source?

    (a) 2.00 A rms.

    (b) 2.40 A rms.

    (c) 1.33 A rms.

    (d) 0.800 A rms.

CHAPTER 8

Gravitation

In this chapter, we'll see how gravity governs the behavior of matter. We'll learn how the potential energy of an object varies as its altitude changes. Then we'll outline Isaac Newton's and Johannes Kepler's conclusions concerning orbital motion and gravitational force. We'll also see how gravitation is related to the geometry of space and time, and what can happen when gravitation becomes extreme.

What Is Gravitation?

In order to explain the acceleration of falling objects, the orbit of the moon around the earth, and the orbits of the planets around the sun, Isaac Newton hypothesized in the 1600s that every particle or object in the universe attracts every other particle or object in a mathematically definable way. This force became known as *gravitation*, also called the *force of gravity* or simply gravity.

THE APPLE INCIDENT

According to a popular legend, Newton got the idea for his *law of gravitation* while sitting in an apple orchard and watching an apple fall to the ground. The details of the event are not universally agreed-on, but we can imagine that he gazed at an apple dangling from a tree branch at the moment it snapped off and fell. Initially its speed was zero, but it immediately began to move, slowly at first, then faster and faster.

Newton might have reasoned that the apple was impelled toward the center of the earth by a relentless, unchanging force. Perhaps he remembered the story about the Italian astronomer Galileo Galilei's experiment involving objects of different masses dropped from the Leaning Tower of Pisa. Newton had studied the work of Johannes Kepler who demonstrated three important rules concerning the orbits of the planets in the solar system. (We'll examine these rules later in this chapter.) Newton knew that force, mass, and acceleration had to be related so that, no matter if the apple on the tree was small or large, its rate of descent would be the same, neglecting air resistance.

Newton, like scientists before him, saw that the earth "pulls" on everything on or near its surface. He must have imagined that the earth's force of gravitation is not limited to the surface, but extends into the sky, beyond the atmosphere, into space, and past the moon, the sun, and the other planets. Of course, Newton could not travel into outer space to do experiments, but he could deduce certain facts from the orbital trajectories and periods of the planets in relation to their distances from the sun. Johannes Kepler had already compiled and analyzed that data in some detail.

ACCELERATION CAUSED BY GRAVITY

Historians may question the truth of the legend of Galileo's "tower experiment" as well as that of Newton's "apple revelation." In any case, a few centuries ago, some curious person conducted a controlled experiment in which two dense objects of differing mass were dropped from a height to find out which one would fall faster—and publicized the results.

Figure 8-1 illustrates a situation in which two objects of mass m_1 and m_2 are dropped at the same time t_0 and allowed to fall freely. Suppose that $m_1 > m_2$. Where will the masses be, relative to each other, at some instant t_1 later than t_0, and at some instant t_2 later than t_1? From elementary mechanics, we know that the magnitude of the force vector on an object is equal to the mass of the object times the magnitude of the acceleration vector on that object. In the scenario of Fig. 8-1, if

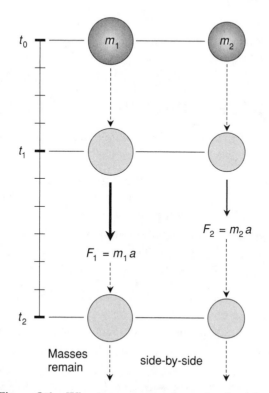

Figure 8-1   When masses m_1 and m_2 are released next to each other at time t_0 and allowed to fall in the absence of air resistance, they remain side-by-side after arbitrary elapsed times t_1 and t_2, even if the gravitational force magnitude F_1 on m_1 differs from the gravitational force magnitude F_2 on m_2.

F_1 represents the magnitude of the *gravitational force* on m_1, F_2 represents the magnitude of the gravitational force on m_2, and a represents the acceleration caused by gravity, then:

$$F_1 = m_1 a = 9.807\ m_1$$
$$F_2 = m_2 a = 9.807\ m_2$$

Here, we consider a to be a constant with a value of 9.807 m/s². We are told that $m_1 > m_2$, so it follows from the above equations that $F_1 > F_2$.

At first thought, it might seem that m_1 would fall faster than m_2 because the force on m_1 is greater than the force on m_2. But the two objects fall at the same rate because m_1 has more *inertia* than m_2. The larger mass resists the acceleration more than the smaller mass does. The effect of variable gravitational force and the effect of variable inertia offset each other.

NEWTON'S LAW OF GRAVITATION

Here is the essence of the hypothesis at which Isaac Newton arrived, and which has become known as *Newton's law of gravitation*.

- Any two particles or objects in the universe, no matter how far apart or how small, exhibit a mutual attractive force of gravitation that can operate through any medium, including empty space.
- The magnitude of the force of gravitation between two particles or objects is directly proportional to the product of their masses and inversely proportional to the square of the distance between their centers of mass.

Now we'll state the law mathematically. Let m_1 and m_2 be the masses of two particles or objects whose centers of mass are separated by distance s. The gravitational force vector **F** acts to pull the objects or particles directly toward each other along a straight line connecting their centers of mass. The magnitude F of this vector is:

$$F = Gm_1m_2 / s^2$$

where F is in newtons, G is a constant that we'll define shortly, m_1 and m_2 are in kilograms, and s is in meters.

Figure 8-2 illustrates how Newton's law of gravitation operates for two objects near each other. Suppose there are no other objects nearby to exert appreciable gravitational force on either object. At A, the objects, with mass m_1 and m_2, are separated by distance s, resulting in gravitational force magnitude F. At B, the mass of the object on the left is doubled but the distance between the mass centers remains the same. The resulting force magnitude doubles to $2F$. At C, the masses of the objects are the same as at A, but the mass centers are twice as far apart. The force magnitude in this case is $F / 4$.

THE GRAVITATIONAL CONSTANT

The constant G in the above equation is called the *gravitational constant* or the *Newtonian constant of gravitation*. It is the same for all particles and objects regardless of size or mass, at least in the vicinity of the earth. Most physicists believe that G has the same value throughout the solar system, as well as everywhere in our galaxy and even beyond our galaxy. But we cannot be absolutely

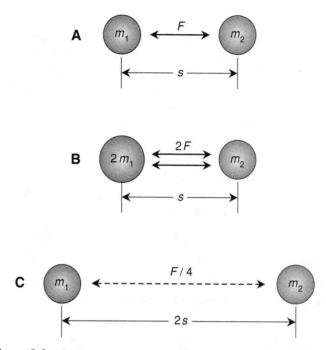

Figure 8-2   At A, two masses m_1 and m_2, separated by distance s, have mutual gravitational force F. At B, one mass is doubled but the distance remains the same; the gravitational force is doubled. At C, the distance is doubled but the masses remain the same; the gravitational force becomes 1/4 as great.

certain that G has the same value at great distances that visually appear to us in the distant past because of the finiteness of the speed of light. Was G the same, say, 10 billion years (10^{10} yr) ago as it is today? According to the *big bang theory* of cosmic evolution, the universe was much smaller 10^{10} yr ago than it is now. Could G be somehow dependent on the size or density of the universe? No one is sure. But the value of G in our part of the universe at the present time has been determined by experimentation. As of the year 2002, the NIST has specified:

$$G = 6.6742 \times 10^{-11} \text{ N} \cdot \text{m}^2/\text{kg}^2$$
$$\pm 0.0010 \times 10^{-11} \text{ N} \cdot \text{m}^2/\text{kg}^2$$

The plus-or-minus symbol indicates the degree of uncertainty. The value most often cited in texts is:

$$G = 6.67 \times 10^{-11} \text{ N} \cdot \text{m}^2/\text{kg}^2$$

The first controlled experiment to find the value of G was conducted in 1798 by a physicist named Henry Cavendish. His apparatus consisted of two objects having precisely known masses m_1 and m_2, whose distance from each other could be varied at will and accurately measured. To determine the force of attraction, a sensitive *torsion balance* was used along with a *light-beam lever* of the sort shown in Fig. 8-3. The balance was allowed to come to rest with m_1 far removed from m_2. Then m_2 was brought close to m_1, and the displacement measured using the light-beam lever. Knowing the amount of force necessary to cause a certain amount of twisting in the fiber, as well as the length of the supporting shaft and the separation between the centers of mass, Cavendish figured out how much attractive force was taking place between the two masses. Then he inferred the value of G by working Newton's formula backwards:

$$F = Gm_1m_2 / s^2$$
$$G = s^2F / (m_1m_2)$$

Cavendish had to be sure that there was no electrical charge on either object, because even the tiniest charge would cause gross errors in the measured value of G.

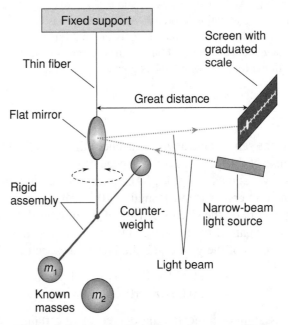

Figure 8-3 Late in the eighteenth century, Henry Cavendish used a device like this to measure the value of the gravitational constant.

A small amount of air movement could have an effect as well. The objects had to be non-ferromagnetic so there would be no magnetic attraction or repulsion. Despite all this potential for error, the value of G found by Cavendish was not much different from the values obtained more recently using far better equipment.

WEIGHT VS. MASS

Although they are often confused by laypeople, *weight* and *mass* are two entirely different things. A 1-kg object possesses the same mass no matter where it is located. Mass is a literal expression of how much matter an object contains. Weight, in contrast, is the force exerted by gravitation or acceleration on an object having a given mass. This force can vary greatly depending on the location of the object. An 80-kg human being weighs more on the earth than on the moon, for example. The Apollo astronauts could vouch for this by direct experience.

Technically, the standard unit of weight is the newton (N), not the kilogram (kg) or the pound (lb). On the surface of the earth we often "translate" newtons into pounds or kilograms using a bathroom scale, but this is an inappropriate conversion because mass and weight are conceptually different. The formal definition of the pound gives it a value of approximately 0.4536 kg, so it's a unit of mass, not weight. An old-time unit called the *poundal* represents weight, but we won't get into poundals here because they're hardly ever mentioned nowadays. One poundal is the force that causes a mass of one pound to accelerate at one foot per second squared.

If your mass is 80.00 kg, then it is 80.00 / 0.4536 = 176.4 lb. If your mass is 122.7 lb, then it is 122.7 × 0.4536 = 55.66 kg. But your weight, F, must be determined using the formula for force in terms of mass m and acceleration a:

$$F = ma$$

where F is in newtons, m is in kilograms, and a is in meters per second squared. If we take the acceleration caused by gravity on the earth's surface to be 9.807 m/s^2, then an 80.00-kg person standing in your living room has a weight of:

$$F = 80.00 \times 9.807$$
$$= 784.6 \text{ N}$$

PROBLEM 8-1

Given that the mass of the earth is approximately 6.0 × 10^{24} kg and the distance from the earth to the moon is about 4.0 × 10^5 km, how hard does the earth's gravitation "pull" on an astronaut whose mass is 80 kg while that astronaut stands on the moon? Determine the answer to two significant figures.

SOLUTION 8-1

First, we must get our units in order. The masses are expressed correctly (in kilograms), but the distance between the earth and the moon must be converted from kilometers to meters. That means we must multiply the figure given by 1000, obtaining 4.0×10^8. If we let m_1 represent the mass of the earth, m_2 represent the mass of the astronaut, and s be the distance between the earth and the moon, then we have:

$$m_1 = 6.0 \times 10^{24}$$
$$m_2 = 80$$
$$s = 4.0 \times 10^8$$
$$G = 6.67 \times 10^{-11}$$

The gravitational force F, in newtons, with which the earth "pulls" on the astronaut is:

$$F = Gm_1m_2 \ / \ s^2$$
$$= 6.67 \times 10^{-11} \times 6.0 \times 10^{24} \times 80 \ / \ (4.0 \times 10^8)^2$$
$$= (3.2 \times 10^{16}) \ / \ (1.6 \times 10^{17})$$
$$= 0.20 \text{ N}$$

The astronaut is not consciously aware of any gravitational force from the earth because the moon "pulls" much harder on the astronaut's body than does the earth, and also because the astronaut orbits the earth along with the moon. But the force is there, nevertheless.

PROBLEM 8-2

Suppose that the moon suddenly stops in its orbit and begins falling toward the earth. What is the initial magnitude of its acceleration vector towards the earth? You can use the data from the previous problem to figure this out. Express the answer to two significant figures. Neglect any effect that might occur as a result of the sun's gravitation.

SOLUTION 8-2

If the moon were to stop moving in its orbit, it would accelerate toward the earth at the same rate as any object on it, including the astronaut. We have found that the force on the astronaut, caused by the earth's gravitation, is 0.20 N. We can figure out the acceleration with which the astronaut falls using the familiar formula from classical mechanics:

$$F = ma$$

where F is the force in newtons exerted on the astronaut by the earth's gravitation, m is the mass of the astronaut in kilograms, and a is the acceleration in meters per second squared with which the astronaut (and therefore the moon and anything else on it) would fall toward the earth. "Plugging in the numbers" and solving for a, we get:

$$0.20 = 80a$$
$$a = 0.20 / 80$$
$$= 0.0025 \text{ m/s}^2$$

PROBLEM 8-3

Plot a graph showing the earth's gravitational force F, in newtons, on a 1.00-kg mass as a function of the distance s from the center of the earth, for values of s ranging from somewhat above the surface to an altitude of 10^8 m. For mathematical simplicity, consider the mass of the earth to be 6.00×10^{24} kg, concentrated entirely at the gravitational center.

SOLUTION 8-3

Imagine a rectangular coordinate system whose horizontal axis is graduated in increments of 10^7 m and whose vertical axis is graduated in increments of 0.1 N. If we can find one point on this graph, we can extrapolate several others and then draw the contour of the plot. To get our initial point P, set $s = 5.00 \times 10^7$ m. If m_1 is the mass of the earth, m_2 is the mass of our object, s is the distance between our object and the center of the earth, and G is the gravitational constant, then for point P, we have these values:

$$m_1 = 6.00 \times 10^{24}$$
$$m_2 = 1.00$$
$$s = 5.00 \times 10^7$$
$$G = 6.67 \times 10^{-11}$$

The gravitational force F at point P is:

$$F = Gm_1m_2 / s^2$$
$$= 6.67 \times 10^{-11} \times 6.00 \times 10^{24} \times 1.00 / (5.00 \times 10^7)^2$$
$$= (4.00 \times 10^{14}) / (2.50 \times 10^{15})$$
$$= 0.160 \text{ N}$$

Newton's law of gravitation tells us that the earth's gravitational force on any object is inversely proportional to the square of the distance between that object and the earth's center of mass. Therefore on the basis of P:

- If we divide s by 2, we must multiply F by $2^2 = 4$.
- If we divide s by 3, we must multiply F by $3^2 = 9$.
- If we multiply s by 2, we must divide F by $2^2 = 4$.

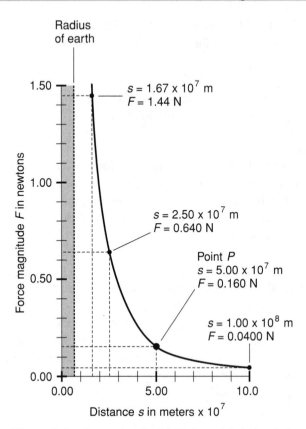

Radius
of earth

$s = 1.67 \times 10^7$ m
$F = 1.44$ N

$s = 2.50 \times 10^7$ m
$F = 0.640$ N

Point P
$s = 5.00 \times 10^7$ m
$F = 0.160$ N

$s = 1.00 \times 10^8$ m
$F = 0.0400$ N

Force magnitude F in newtons

Distance s in meters x 10^7

Figure 8-4   Illustration for Problem and Solution 8-3.

The plot, derived on the basis of our initial point P and three additional points resulting from the above operations, is shown in Fig. 8-4.

Gravitational Fields

The concept of a *gravitational field* is abstract. We can think of it as a set of vectors, known as a *vector field*, surrounding any object having mass. Then all the vectors in the gravitational field produced by an object point toward the center of mass of that object. The magnitude of any given vector depends on the distance of its originating point from the center of mass of the object; this magnitude is inversely proportional to the square of the distance.

GRAVITATIONAL FIELD STRENGTH

The magnitude of the gravitational field vector produced by a massive object Y at any point P in space can be found by placing a small object X of mass m_x at P, determining the gravitational force vector magnitude F_{xy} between X and Y according to Newton's law of gravitation, and then dividing F_{xy} by m_x. This ratio represents the *gravitational field strength* produced by object Y at point P.

Gravitational field strength is expressed in meters per second squared, just like acceleration. In fact, gravitational field strength is quantitatively identical to the acceleration caused by gravity. Because of this, gravitational field strength is often called *gravitational acceleration* or the *acceleration of gravity*. It represents the actual rate of acceleration with which a small object X, located at a specific point P, will fall toward the center of mass of a much more massive object Y. In the absence of external gravitational influences, the gravitational field strength at any point near a massive object depends only on the mass of the object and the distance between the point and the center of mass of that object.

When all the points in space near a massive object are taken into consideration, a *gravitational field-strength function* can be defined. It is a *vector function* with a domain having three *independent variables*, representing the coordinates of each point surrounding the object. Examples of such variables might be x, y, and z (east-west, up-down, and south-north) in a rectangular system of coordinates near a particle in a laboratory, or r, θ, and ϕ (radial distance from the earth's center, celestial latitude, and celestial longitude) in space near our planet. The range of the vector function has two *dependent variables*: one representing the gravitational force vector direction, denoted dir **F**, and the other representing the gravitational field-strength magnitude, denoted $|\mathbf{F}|$ or F.

POINT AND NON-POINT MASSES

Newton's law, as we have described it above, applies only to *point masses*. Those are objects that can be considered, for the purpose of calculating gravitational effects, as having all their matter contained in a single point. In some situations, real-life objects behave as point masses, but often they do not. When we have a *non-point mass*, we cannot directly apply Newton's law of gravitation to the object as a whole. Instead, we must individually consider the effects of many small bits or *parcels* of the whole, considering each parcel as a point mass. These effects add together to create the net gravitational force. But even then, we can apply Newton's law with confidence to an object only when that object is solid throughout.

Imagine a large solid object as many parcels of matter, each producing its own gravitational field. When we're at or above the surface of a perfectly spherical solid object of uniform density, the average position of all these parcels corresponds to the center of the sphere. If we integrate to find the net combined gravitational effect of all the parcels, we come up with the same gravitational force vector **F** as would exist if the entire mass of the sphere were concentrated at the center. This integration process can also be done for irregular objects such as asteroids, or for objects that do not have uniform density throughout. In those situations, the integration process is more complicated than it is for a sphere of uniform density. Nevertheless, there always exists a point called the *center of gravity* that defines a location for implementing Newton's formula if we stay far enough above the surface of an object, no matter how irregular or non-uniform the object might be. But *we must stay outside the object.*

THE LAW OF SUPERPOSITION

When a point mass X is influenced by two or more other point masses, the net gravitational force vector on X, caused by all the other point masses collectively, is equal to the vector sum of the force vectors produced by each of the other point masses individually. This is called the *law of superposition* or the *superposition theorem* for gravitational force vectors.

Figure 8-5 illustrates the law of superposition in a simple case where a point mass X is influenced by two other point masses Y and Z. In drawing A, vector

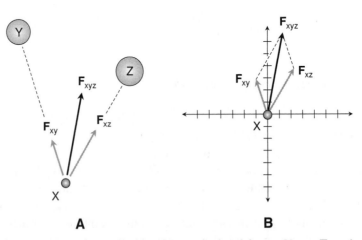

A **B**

Figure 8-5 The law of superposition for gravitational forces. Vector $\mathbf{F}_{xyz}$, the net force imposed on X by Y and Z together, is equal to the sum of $\mathbf{F}_{xy}$, the force imposed on X by Y, and $\mathbf{F}_{xz}$, the force imposed on X by Z. Drawing A shows a hypothetical example. Drawing B shows the vector summation geometry for that example.

$\mathbf{F}_{xy}$, representing the force produced on X by Y, "pulls" X toward Y. Vector $\mathbf{F}_{xz}$, representing the force produced on X by Z, "pulls" X toward Z. The net gravitational force vector on X, resulting from the combined gravitational forces imposed by Y and Z, is shown as vector $\mathbf{F}_{xyz}$.

Vector $\mathbf{F}_{xyz}$ can be found in terms of $\mathbf{F}_{xy}$ and $\mathbf{F}_{xz}$ by means of vector addition. This is shown geometrically in Fig. 8-5B. To express it as a vector equation, we write simply:

$$\mathbf{F}_{xyz} = \mathbf{F}_{xy} + \mathbf{F}_{xz}$$

This formula can be extrapolated to any number of vectors in a sum. Therefore, we can determine the net gravitational force vector on a particular point mass no matter how many other point masses happen to be in the vicinity, as long as we know the magnitudes and directions of the force vectors produced by each point mass. Suppose we are given coordinates for many point masses and the actual mass of each, and we want to find the net gravitational force vector imposed on any one point mass X by all the others. We can use Newton's law of gravitation to find the gravitational force vectors on X caused by each of the other point masses. Then we can invoke the law of superposition, adding up all the individual force vectors.

The law of superposition for gravitational force vectors applies only in the simplistic sense when X is a point mass. That means the radius of X must be very small compared with the distance between it and other objects that exert gravitational forces on it. In addition, the object X in examples such as that of Fig. 8-5 must consist of solid matter. The law of superposition does not apply to the combined gravitational forces from the sun and the moon that govern the earth's oceanic tides, for example, because the ocean cannot be considered as a point mass of solid matter.

GRAVITATION INSIDE A SOLID SPHERE OF UNIFORM DENSITY

Once you venture inside a large spherical object such as the earth, Newton's law can no longer be applied as if the object is a point mass. If you were to go a long way underground, some of the earth's mass would be beneath you, but some would be above you. Each individual parcel of matter above you would impart a force "pulling" you away from the center. In fact, you would be "pulled" to some extent in every possible direction at once! For this reason, you would weigh less inside the earth than you do at the surface. The deeper you went, the greater the effect would become until, if you carved out a bunker at the earth's center of mass, you would experience a net gravitational force of zero. You would be "weightless," just as if you were floating in interplanetary space. At points between the center and the surface, the magnitude of the gravitational force vector would vary with the distance from the center, but the vector would always point straight toward the center. Force vectors in all other directions would cancel each other out.

Suppose you descend beneath the surface of a solid sphere X_r of uniform density having radius r. Imagine that you go down until you're at a distance s from the center of X_r, where $s < r$. The gravitational force you experience corresponds to the force that would exist if you were standing on the surface of an imaginary sphere X_s having radius s, and having the same uniform density and the same center of mass as the larger sphere X_r (Fig. 8-6). You can ignore the gravitational force produced by the "shell" with inner radius s and outer radius r. All the parcels of

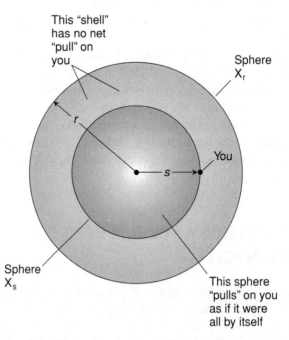

Figure 8-6 If you descend beneath the surface of a spherical mass of uniform density, the gravitational force you experience is less than the force you experience on the outer surface. Illustration for Problem and Solution 8-4.

matter that make up such a "shell," taken together, contribute a net gravitational force of zero on any point inside the "shell" or on its inner surface. This can be proven by mathematical integration. We won't go through that demonstration, but let's work out a problem to illustrate an example.

PROBLEM 8-4

Consider the mass m_e of the earth to be 6.0×10^{24} kg, its radius r_e to be 6.4×10^6 m, and the acceleration of gravity at the surface to be 10 m/s². Suppose the earth has perfectly uniform density everywhere inside. What is the acceleration of gravity on an object S halfway between the center of the earth and the surface? Round the answer off to two significant figures.

SOLUTION 8-4

Because we assume that the matter inside the earth is uniformly distributed, we can also assume that the mass contained within any "subsphere" of the earth is directly proportional to the ratio of the volume of that "subsphere" to the volume of the whole planet. From your solid geometry courses, you should remember that the volume of a sphere is directly proportional to the cube of its radius. Our object S is halfway between the center of the earth and the surface, so S is on the "surface" of a "subsphere" X having half the earth's radius. That means "subsphere" X has $(1/2)^3$, or 1/8, of the earth's volume, and hence 1/8 of the earth's mass. If we let m_x be the mass of "subsphere" X, then:

$$m_x = m_e / 8$$
$$= 6.0 \times 10^{24} / 8.0$$
$$= 7.5 \times 10^{23} \text{ kg}$$

Let's assign our object S an arbitrary mass of $m_s = 1.0$ kg. (The reason for this choice will soon be apparent!) We have been told that the distance r_s of object S from the earth's center is equal to half of the earth's radius r_e:

$$r_s = r_e / 2$$
$$= 6.4 \times 10^6 / 2.0$$
$$= 3.2 \times 10^6 \text{ m}$$

We can use Newton's law of gravitation to determine the gravitational force vector magnitude F_s on our object S:

$$F_s = G m_x m_s / r_s^2$$
$$= 6.67 \times 10^{-11} \times 7.5 \times 10^{23} \times 1.0 / (3.2 \times 10^6)^2$$
$$= (5.0 \times 10^{13}) / (1.0 \times 10^{13})$$
$$= 5.0 \text{ N}$$

Dividing this by the mass m_s of our object S, which we have conveniently chosen to be 1.0 kg, gives us a gravitational acceleration magnitude $a_g = 5.0$ m/s^2. That's half the gravitational acceleration magnitude at the surface.

The above result might make you think that, inside any solid spherical planet having uniform density throughout, the gravitational acceleration varies in a linear manner with the distance from the center. If you have that idea, you're right! (Would you like to prove it as an optional exercise?)

POTENTIAL ENERGY VS. ALTITUDE

In Chapter 2, we derived a formula for the potential energy gained by an object that moves straight upward in a gravitational field having constant intensity. Let's review that briefly now. Then we'll consider what happens when an object rises over such a long distance that the gravitational field strength diminishes during the journey.

Suppose a force of magnitude F is applied to an object directly against the force of gravity, so the object moves straight upward from altitude h_1 to altitude h_2. If the gravitational field strength remains constant over the entire upward distance, then so does F, so the potential energy, E_p, that the object gains is given by:

$$E_p = F(h_2 - h_1)$$

where E_p is in joules, F is in newtons, and h_1 and h_2 are in meters.

We can use the old standby formula, $F = ma$, to get another expression for the increase in potential energy acquired by an object of mass m that gains altitude in a gravitational field whose strength is defined by the acceleration a_g:

$$E_p = ma_g(h_2 - h_1)$$

where E_p is in joules, m is in kilograms, a_g is in meters per second squared, and h_1 and h_2 are in meters.

Now consider a situation in which an object is pushed (by a rocket engine, say) upward so far that the gravitational field strength, and hence the gravitational force on it, decreases during the journey. Let s be the instantaneous separation between the centers of mass of the planet and the ascending object. Let s_1 be the initial separation, and let s_2 be the final separation. Let $F(s)$ be a function that represents the force F in terms of s. In this case, the potential energy E_p that the movable object gains during the upward journey is given by:

$$E_p = \int_{s_1}^{s_2} F(s)\, ds$$

where E_p is in joules, F is in newtons, and s, s_1, and s_2 are in meters. If the object does not travel straight upward during the whole ascent, this equation still holds as long as we consider s, s_1, and s_2 in terms of radial distances between the centers of mass of the planet and the object. Only the radial or upward component of the motion contributes to a gain in potential energy; lateral or sideways components do not.

Let m_1 be the mass of our planet, in whose gravitational field we toil. Suppose a rising object in the earth's gravitational field has mass m_2. Then we know the following for any specific separation s_o between the centers of mass of the earth and the object:

$$F = Gm_1 m_2 / s_o{}^2$$

We can now restate the function $F(s)$ as follows, substituting the variable separation s for the specific separation s_o:

$$F(s) = Gm_1 m_2 / s^2$$

By substitution in the preceding integral equation, and by working the integral out, we obtain the following:

$$
\begin{aligned}
E_p &= \int_{s_1}^{s_2} (Gm_1 m_2 / s^2)\, ds \\
&= Gm_1 m_2 \int_{s_1}^{s_2} (1/s^2)\, ds \\
&= Gm_1 m_2 (-1/s) \Big|_{s_1}^{s_2} \\
&= Gm_1 m_2 [-1/s_2 - (-1/s_1)] \\
&= Gm_1 m_2 (1/s_1 - 1/s_2)
\end{aligned}
$$

In case you've forgotten the *evaluation notation* (or if you've never seen it before), the vertical bar with s_1 at lower right and s_2 at upper right means that the expression to its left is to be evaluated from $s = s_1$ to $s = s_2$.

The above formula represents the amount by which the potential energy of a small object increases as it rises over a great distance in the gravitational field of a much larger body. The potential energy thus defined is relative, not absolute. Many physics texts consider the absolute gravitational potential energy of a small object to be zero at a distance of "infinity" from a vastly larger body, and to become more and more negative as it descends. A rising object still gains gravitational potential energy in this model because its absolute gravitational potential energy gets less and less negative. Physicists may make this strange choice of a reference point—"infinity," which does not exist in the real universe and is a subject of controversy even among pure mathematicians—because it produces the least messy mathematics for calculating the kinetic and mechanical energies of orbiting objects.

The discussion here is based on the assumption that the small object does not have significant gravitational effect on the larger one. An example is an artificial satellite in the earth's gravitational field, or a planet in the sun's gravitational field. Systems in which the masses are not vastly different, such as the earth-moon system or a double star, are more complicated.

PROBLEM 8-5

Calculate the amount of potential energy gained by a 2.5-kg object that is pushed straight upward from the earth's surface to an altitude of 2900 km above the surface (not above the center!). Consider the mass of the earth to be 6.0×10^{24} kg, and consider its radius to be 6.4×10^6 m. Assume that the earth has perfectly uniform density everywhere inside so it can be treated as a point mass located at its center. Neglect any possible gravitational influences from the sun or the moon. Express the answer to two significant figures.

SOLUTION 8-5

We are given the mass of the earth and the mass of our object, and we know the gravitational constant. Therefore, let's set variables as follows:

$$m_1 = 6.0 \times 10^{24}$$
$$m_2 = 2.5$$
$$G = 6.67 \times 10^{-11}$$

We are told the object starts at the surface, so we can set:

$$s_1 = 6.4 \times 10^6$$

The object rises 2900 km above the surface, which converts to 2.9×10^6 m. Therefore, we can set:

$$s_2 = 6.4 \times 10^6 + 2.9 \times 10^6$$
$$= 9.3 \times 10^6$$

Deriving E_p, the gain in potential energy, is simply a matter of plugging these numbers into the formula we derived a while ago:

$$
\begin{aligned}
E_p &= Gm_1m_2 \, (1/s_1 - 1/s_2) \\
&= 6.67 \times 10^{-11} \times 6.0 \times 10^{24} \times 2.5 \times [1/(6.4 \times 10^6) - 1/(9.3 \times 10^6)] \\
&= 1.0 \times 10^{15} \times (1.5625 \times 10^{-7} - 1.0753 \times 10^{-7}) \\
&= 1.0 \times 10^{15} \times 4.872 \times 10^{-8} \\
&= 4.9 \times 10^7 \text{ J}
\end{aligned}
$$

MINIMUM ESCAPE SPEED

Everyone has heard the expression *escape velocity* in reference to spacecraft destined for the moon or planets. Many people have memorized the phrase "twenty-five thousand miles an hour" (which translates to about 4.0×10^4 km/h) for the minimum speed with which an object would have to be hurled straight upward from the surface of the earth in order to fly off into interplanetary space.

As you know, velocity is a vector quantity with direction as well as speed, so things aren't quite as simple as the timeworn phrase suggests. If the direction of the escape velocity vector is straight away from the center of the earth, then its minimum magnitude is indeed approximately 4.0×10^4 km/h or 4.0×10^7 m/h. To get the figure in terms of distance per second, we can divide by 3600, obtaining approximately 11 km/s or 1.1×10^4 m/s. This is the *minimum escape speed*, which we will call $v_{\text{min-esc}}$, for an object on the earth's surface, neglecting gravitational effects of the moon and sun, and also neglecting atmospheric friction.

We can derive $v_{\text{min-esc}}$ for the earth's surface based on what we have learned so far about potential and kinetic energy. Imagine a huge slingshot capable of propelling stones at enormous speed. You put a stone having mass m_2 in this slingshot and aim it straight up. If the stone is released with initial speed v, then its initial kinetic energy E_{ki} is given by:

$$E_{\text{ki}} = m_2 v^2 / 2$$

where E_{ki} is in joules, m_2 is in kilograms, and v is in meters per second. (You learned this in Chapter 2.) After the stone is released, it slows down because of gravitation. Because the stone loses speed, it loses kinetic energy. But it gains potential energy as its altitude increases. The *total mechanical energy*, which is the sum of the potential and kinetic energy, stays constant. Unless the stone is moving at escape speed, it eventually comes to rest at some instant and then starts to fall. At the instant when the stone is at rest after having traveled upward, all of its initial kinetic energy has been translated into a gain in potential energy. When the stone starts to fall, this potential energy starts changing back to kinetic energy. When the stone finally hits the surface at the end of its journey, it has the same kinetic energy, E_{ki}, that it had when it came out of the slingshot.

Now suppose that a stone having mass m_2 is hurled straight upward with just enough speed to prevent it from ever falling back to earth again, but no faster than that. We can theoretically say, neglecting the presence of other objects in the universe, that in this case the stone will come to rest "at infinity." Let s_1 be the starting distance of the stone from the center of the earth (the radius of the earth). Let s_2 be the distance of the stone from the center of the earth when the stone comes to rest ("infinity"). We have:

$$s_1 = 6.4 \times 10^6$$
$$s_2 = \infty$$

The sideways-8 symbol represents "infinity." Pure mathematicians might cringe now, knowing that we're evaluating a fraction with an infinite denominator! But let's proceed according to the hypothesis that any finite number divided by "infinity" is equal to zero.

As the stone rises from the surface and finally comes to rest "at infinity," we can calculate how much potential energy it gains using the formula we derived earlier. Let m_1 be the earth's mass, which is 6.0×10^{24} kg. We aren't told m_2, the mass of the stone, but as things work out, it doesn't matter. We have:

$$E_p = Gm_1m_2 \, (1/s_1 - 1/s_2)$$
$$= 6.67 \times 10^{-11} \times 6.0 \times 10^{24} \times m_2 \times [1/(6.4 \times 10^6) - 1/\infty]$$

Letting $1/\infty = 0$, we have:

$$E_p = 6.67 \times 10^{-11} \times 6.0 \times 10^{24} \times m_2 \times 1/(6.4 \times 10^6)$$
$$= (6.2531 \times 10^7) \, m_2$$

Let's leave extra digits in the calculation until we're done, and then we'll round off to two significant figures (the extent of the accuracy of our input data) at the end of the derivation. That will minimize cumulative rounding errors.

Now imagine that the stone has come to rest after coasting upward for an infinitely great distance. Because it has stopped moving, its kinetic energy is zero. All the potential energy it has gained during this journey is the same as the kinetic energy E_{ki} it originally had at the earth's surface when it came out of the slingshot. If v was its initial velocity, that means:

$$E_p = m_2v^2 / 2$$

Substituting for E_p from above and solving for v gives us:

$$(6.2531 \times 10^7) \, m_2 = m_2v^2 / 2$$
$$6.2531 \times 10^7 = v^2 / 2$$
$$v^2 = 1.2506 \times 10^8$$
$$v = 1.1 \times 10^4 \text{ m/s}$$

That's the minimum escape speed $v_{\text{min-esc}}$ for any object at the surface of the earth. Three facts are worth noting here:

- The value of $v_{\text{min-esc}}$ for an object does not depend on the mass of that object, as long as all other factors are held constant.
- The value of $v_{\text{min-esc}}$ for an object depends on the mass of the planet or other large body from which it must escape.
- The value of $v_{\text{min-esc}}$ for an object depends on the initial altitude or distance of the object from the center of mass of the planet or other large body from which it must escape.

Orbits and Kepler's Laws

Every object that orbits the earth, the sun, or any other massive celestial body is in a state equivalent to *free fall*. The same is true of objects in orbit around each other. The moon "falls" around the earth, Jupiter's moons "fall" around that planet, and double stars "fall" around one another—endlessly, at least in theory.

CENTRIPETAL ACCELERATION

An object X orbiting a much more massive body Y constantly accelerates because of the force of gravitation imposed by Y. This acceleration, which can be manifest as a change in the speed of X, a change in the direction of X, or both, is called *centripetal acceleration*. Let's symbolize the centripetal acceleration vector as $\mathbf{a}_c$, and its magnitude as a_c.

Suppose an object X is in orbit around a much more massive body Y, and the gravitational force from Y is the only force acting on X. In this situation, dir $\mathbf{a}_c$ is always toward the center of mass of Y, just as is dir $\mathbf{a}_g$, the direction of the gravitational acceleration vector. Also, a_c is always the same as a_g, the magnitude of the gravitational acceleration caused by Y. The following simple rule holds at every instant in time:

$$\mathbf{a}_c = \mathbf{a}_g$$

Neither the vector magnitudes nor their directions depend on the mass of the orbiting object X. But the vector magnitudes do depend on the distance between the centers of mass of objects X and Y. If, during the course of an orbital revolution, the instantaneous distance s between the centers of mass of X and Y remains constant, then the instantaneous values of a_c and a_g also remain constant. But if s varies, then the instantaneous values of a_c and a_g vary inversely in proportion to the square of s from moment to moment. The value of s is called the *radius* of the orbit.

KEPLER'S LAWS

Johannes Kepler published his rules relating to planetary motions in the early part of the seventeenth century. Much of Newton's work derived from the pioneering efforts of Kepler. These three rules, which have become known as *Kepler's laws*, can be stated as follows.

- *Kepler's first law:* Each planet follows an elliptical orbit around the sun, with the sun's center of mass located at one focus of the ellipse.

- *Kepler's second law:* An imaginary straight line connecting the center of mass of any planet with the center of mass of the sun sweeps out equal areas in equal periods of time.

- *Kepler's third law:* For any two planets, the ratio of the squares of their *sidereal periods* (the time it takes to complete one orbit relative to the background of distant stars) is directly proportional to the ratio of the cubes of their orbital radii.

Kepler came up with his three principles and refined them over a time frame of several years. The first two rules were finalized in 1609, and the last one came out in 1618. The first two laws are illustrated in Fig. 8-7, and the third law is illustrated in Fig. 8-8.

Kepler's first two laws can be applied directly to orbiting objects in general, not only the orbits of the planets around the sun. The first two laws work for satellites revolving around the earth or the moon, the moon as it orbits the earth, and the moons of planets such as Jupiter and Saturn. Many such objects follow orbits that are far from perfect circles. Kepler's third law must be modified in such cases. We'll look into that shortly.

Since the time of Galileo, Newton, Kepler, and their colleagues, the science of orbital mechanics has been refined, largely as a result of the work of Einstein in the twentieth century. *Relativistic effects* introduce small errors into Newtonian and Keplerian physics. These discrepancies can be observed and measured. They

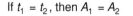

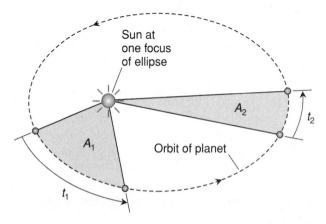

Figure 8-7 Kepler's first and second laws. The planetary orbit is an ellipse with the sun at one focus. Equal areas A_1 and A_2 are swept out in equal periods of time t_1 and t_2.

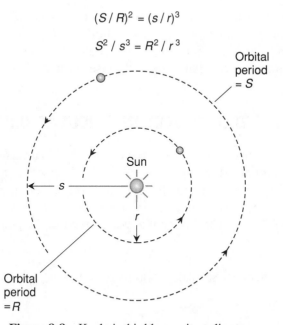

$$(S/R)^2 = (s/r)^3$$

$$S^2/s^3 = R^2/r^3$$

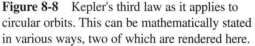

Figure 8-8 Kepler's third law as it applies to circular orbits. This can be mathematically stated in various ways, two of which are rendered here.

are not large, except when gravitational forces become far more extreme than anything encountered in the solar system.

TANGENTIAL SPEED OF OBJECT IN CIRCULAR ORBIT

Kepler's second law logically implies, with the help of the axioms and theorems of Euclidean geometry, that if an object X orbits a much more massive body Y in a perfectly circular orbit, then X travels at constant *tangential orbital speed*. (If you want to rigorously prove this as an exercise, feel free to do it!) Let m_y be the mass of the large central body Y, and let s be the distance between the centers of mass of X and Y. Because the orbit is circular, s is constant. We can find the constant tangential orbital speed v_x for object X using this formula:

$$v_x = (Gm_y/s)^{1/2}$$

where v_x is in meters per second, G is the gravitational constant, m_y is in kilograms, s is in meters, and the 1/2 power represents the positive square root.

Note that the tangential orbital speed of X does not depend on the mass of X. This is why, for example, an astronaut can step out of a space station as it orbits the earth without flying off into a different orbit at blinding speed because her mass is so much less than the mass of the station. But if she jumps out without an "umbilical cord" attaching her to the station, she'd better have a rocket backpack. Otherwise the impulse from her jump will cause her to drift *slowly* away!

ANGULAR SPEED OF OBJECT IN CIRCULAR ORBIT

If some object X orbits a much more massive object Y in a perfectly circular orbit at constant tangential orbital speed, then X travels at constant *angular orbital speed* as well. The relation among the tangential orbital speed v_x of object X, the angular orbital speed ω_x of object X, and the radius s of the orbit is based on the formula for uniform circular motion that you learned in Chapter 3:

$$v_x = s\omega_x$$

where v_x is in meters per second, s is in meters, and ω_x is in radians per second. Conversely:

$$\omega_x = v_x \, / \, s$$

By substitution in the formula for v_x in terms of G, m_y, and s, we obtain:

$$\omega_x = v_x \, / \, s$$
$$= (Gm_y \, / \, s)^{1/2} \, / \, s$$

PERIOD OF OBJECT IN CIRCULAR ORBIT

Once again, imagine that some object X orbits a much more massive body Y in a perfectly circular orbit. We can determine the *orbital period* for X, symbolized T_x, directly from the tangential orbital speed v_x and the orbital radius s, as follows:

$$T_x = 2\pi s \, / \, v_x$$

where T_x is in seconds, s is in meters, and v_x is in meters per second. We can also calculate T_x from the angular orbital speed ω_x and the orbital radius s:

$$T_x = 2\pi \, / \, \omega_x$$

where T_x is in seconds, s is in meters, and ω_x is in radians per second. In terms of the orbital radius s and the mass m_y of the central body Y, we can derive the following formula from either of the two preceding:

$$T_x = 2\pi \, [s^3 \, / \, (Gm_y)]^{1/2}$$

where T_x is in seconds, s is in meters, G is the gravitational constant, and m_y is in kilograms.

ORBITAL BARYCENTER

Until now, we have assumed that the mass of an orbiting object X is negligible compared with the mass of the central body Y around which X revolves. When that is true, the center of the orbit of X almost exactly coincides with the center of mass of Y. An example of such a near-ideal situation is an artificial satellite orbiting the earth. Each planet-sun pair, considered individually, is close to perfect in that respect, too.

When an object X orbits another object Y in a circle and the masses are not vastly different, then neither object's center of mass is very close to the center of the orbit. In the most simplistic case, X and Y orbit each other in concentric circles around a common center of mass known as the *orbital barycenter*. You can think of such a system as a twirling "baton" with X at one end and Y at the other. The "twirler's" hand is at the orbital barycenter, which is closer to the end of the baton where the body with the larger mass is located. The "baton" is a rotating geometric line segment whose length is equal to the distance between the centers of mass of the two bodies.

Suppose two bodies X and Y are in mutual circular orbit and their masses are not radically different. Let m_x be the mass of X, let m_y be the mass of Y, and let s be the distance between their centers of mass. Suppose that $m_y \geq m_x$. Because the orbits are both perfect circles, s is a constant. Let s_{by} be the distance between the orbital barycenter and the center of mass of body Y. Then the following equation holds:

$$s_{by} = m_x \, s \, / \, (m_y + m_x)$$

where s_{by} and s are in meters, and m_x and m_y are in kilograms.

WHEN ORBITS ARE NOT CIRCULAR

In the real universe, an orbit is never a mathematically perfect circle. Every actual orbit in space is an ellipse that is more or less "out of round." The orbit of every known planet in the solar system (except Pluto, if you want to call it a planet) is close enough to circular that good approximations of its tangential speed, angular speed, and period can be obtained by considering its orbit as a perfect circle whose radius is equal to its *mean orbital radius*, which is the orbital radius averaged over one complete revolution around the sun.

When an orbit departs greatly from a circle, as is the case with some asteroids and nearly all observable periodic comets, the formulas from the preceding several paragraphs can no longer be directly applied. In addition, Kepler's third law, as we stated it before, must be refined. Here is a more comprehensive version of Kepler's third law:

- For any two objects X_1 and X_2 in orbit around a vastly more massive central object Y, the square of the ratio of the orbital periods is directly proportional to the cube of the ratio of the *semimajor axes* of the orbits.

When an object X orbits a much more massive body Y in an elliptical orbit, the semimajor axis a_x of the elliptical orbit is half the diameter of the ellipse as measured "the long way." It can be more rigorously defined as the average of the *minimum orbital radius*, symbolized s_{min} (the distance between the centers of mass of X and Y when they are closest to each other) and the *maximum orbital radius*, symbolized s_{max} (the distance between the centers of mass of X and Y when they are farthest apart):

$$a_x = (s_{min} + s_{max}) / 2$$

Under these circumstances, we can determine the orbital period T_x for object X in terms of the length of the semimajor axis a_x of the orbit and the mass m_y of the central body Y, as follows:

$$T_x = 2\pi \, [a_x^3 / (Gm_y)]^{1/2}$$

where T_x is in seconds, a_x is in meters, G is the gravitational constant, and m_y is in kilograms. This formula is the same as the formula for the period of an object in a circular orbit, except that we specify the length of the semimajor axis rather than the orbital radius.

The concepts of tangential and angular speed, and the way in which they are related, become difficult to define and quantify when an orbit is an ellipse that has a moderate or large degree of *eccentricity*. (Eccentricity is an expression of the extent to which an ellipse departs from a circle.) Also, the determination of the barycenter for a pair of mutually orbiting bodies becomes complicated when either orbit is an ellipse having significant eccentricity, or when there are three or more objects in mutual orbit. The most complicated scenarios are those in which many objects orbit in a "swarm" around a central point in space. An example of such a system is our Milky Way galaxy.

PROBLEM 8-6

Assume that the distance s_{me} between the centers of mass of the earth and the moon is a constant 4.0×10^8 m, the mass m_e of the earth is 6.0×10^{24} kg, and the mass

m_m of the moon is 7.4×10^{22} kg. Also assume that the density of the earth and moon are both uniform throughout, so both objects can be considered as point masses. On the basis of this simplified data, where is the orbital barycenter of the earth-moon system? Ignore any possible gravitational effect from the sun. Express the answer to two significant figures.

SOLUTION 8-6

From the above formula for the orbital barycenter, let $m_y = m_e$, $m_x = m_m$, and $s = s_{me}$. Then we have:

$$
\begin{aligned}
s_{by} &= m_m \, s_{me} / (m_e + m_m) \\
&= (7.4 \times 10^{22} \times 4.0 \times 10^8) / (6.0 \times 10^{24} + 7.4 \times 10^{22}) \\
&= (2.96 \times 10^{31}) / (6.074 \times 10^{24}) \\
&= 4.9 \times 10^6 \text{ m}
\end{aligned}
$$

The radius of the earth is approximately 6.4×10^6 m. Therefore, given the values and conditions stated in this problem, the orbital barycenter of the earth-moon system is inside the earth, about three-quarters of the way from the center to the surface.

PROBLEM 8-7

Imagine Comet X in an elliptical orbit around the sun. Suppose the minimum orbital radius is 0.15 times the earth's mean orbital radius, and the maximum orbital radius is 25 times the earth's mean orbital radius. What is the period of Comet X in earth years? Consider the earth's mean orbital radius to be 1.5×10^8 km. Consider the mass of the sun to be 2.0×10^{30} kg. Consider an earth year (1 yr) to be 365.25 days. Ignore gravitational effects that might be caused by the planets. Express the answer to two significant figures.

SOLUTION 8-7

The most difficult aspect of this problem is getting the variables and units correct and in standard form! First, let's find the length of the semimajor axis of the orbit. The minimum orbital radius of Comet X, s_{min}, is:

$$
\begin{aligned}
s_{min} &= 0.15 \times 1.5 \times 10^8 \\
&= 2.25 \times 10^7 \text{ km} \\
&= 2.25 \times 10^{10} \text{ m}
\end{aligned}
$$

The maximum orbital radius of Comet X, s_{max}, is:

$$
\begin{aligned}
s_{max} &= 25 \times 1.5 \times 10^8 \\
&= 3.75 \times 10^9 \text{ km} \\
&= 3.75 \times 10^{12} \text{ m}
\end{aligned}
$$

The length a_x of the semimajor axis of the orbit of Comet X is therefore:

$$a_x = (s_{min} + s_{max}) / 2$$
$$= (2.25 \times 10^{10} + 3.75 \times 10^{12}) / 2$$
$$= (3.7725 \times 10^{12}) / 2$$
$$= 1.88625 \times 10^{12} \text{ m}$$

We can safely assume that the mass of Comet X is negligible compared with the mass of the sun, so we can apply the formula stated earlier for the period of an elliptical orbit. Let T_x be the orbital period of X in seconds, and let m_y be the mass of the sun. Consider $\pi = 3.14$. Then:

$$T_x = 2\pi [a_x^3 / (Gm_y)]^{1/2}$$
$$= 2.00 \times 3.14 \times [(1.88625 \times 10^{12})^3 / (6.67 \times 10^{-11} \times 2.0 \times 10^{30})]^{1/2}$$
$$= 6.28 \times [(6.71116 \times 10^{36}) / (1.334 \times 10^{20})]^{1/2}$$
$$= 6.28 \times (5.031 \times 10^{16})^{1/2}$$
$$= 6.28 \times 2.243 \times 10^8$$
$$= 1.41 \times 10^9 \text{ s}$$

Finally, we must divide this result by the number of seconds in a year. If we consider 1 yr = 365.25 mean solar days (msd) and 1 msd = 24 h (exactly), then 1 yr = 3.1558×10^7 s. The period $T_{x\text{-yr}}$ of Comet X in earth years is therefore:

$$T_{x\text{-yr}} = (1.41 \times 10^9) / (3.1558 \times 10^7)$$
$$= 45 \text{ yr}$$

As we've mentioned before, the use of extra digits in the calculation process, with rounding done only at the last step, minimizes cumulative rounding errors. The number of extra digits used in some of the above figures is admittedly arbitrary. The only important consideration is that we use enough extra digits, as displayed on our calculators, to be sure we get an accurate figure for rounding when we finish things up.

Extreme Gravitation

According to one theory of stellar evolution, large dying stars are thought to crush themselves under their own gravitation. As they cool off, they contract. The gravitational field strength at the surface increases because of this contraction. Eventually, the minimum escape speed exceeds the speed of light in free space. The ultimate result of all this is a self-contained "mini-universe" known as a *black hole*. In this section, we will examine (in a simplistic sense, because this is an exotic field of physics) what happens when objects undergo *gravitational collapse*.

THE BLACK HOLE

When the gravitational field strength at the surface of a collapsing star becomes great enough, no known force can stop it from contracting to a geometric point that contains all the mass of the star and that has infinite density. Such a point— a true point mass!—is called a *space-time singularity*. It is surrounded by a zone whose outer boundary is known as the *event horizon*. Because the minimum escape speed inside the event horizon is greater than the speed of light in free space, it is tempting to conclude that nothing can get out of a black hole once it falls in. (That is an oversimplification, but further technical detail is beyond the scope of this book.) Figure 8-9 illustrates how the gravitational field strength varies near a space-time singularity.

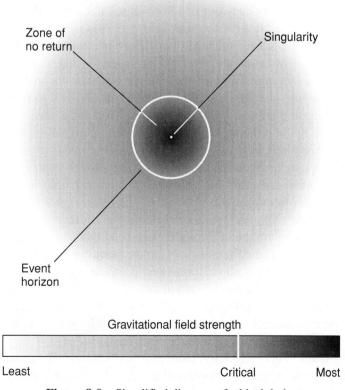

Figure 8-9   Simplified diagram of a black hole.

According to Einstein's *theory of general relativity*, gravitational collapse causes a slowing-down of time relative to the outside universe. This phenomenon is called *gravitational time dilation*. In a black hole, the time-dilation function "blows up," suggesting that time stops altogether as an outside observer would witness it. A person watching a star going into gravitational collapse would see a black sphere having a radius equal to the radius of the event horizon. This is where the expression "black hole" comes from. (Actually, the edge of the sphere might glow faintly because of scattered starlight that has been almost, but not quite, captured and pulled in.) Objects falling into the black hole would seem to slow down as they approached the event horizon, never quite falling through it, but growing dimmer and dimmer as they appeared to "get stuck" to it.

GRAVITY: THE ULTIMATE FORCE

As the gravitational field at the surface of a collapsing star becomes more intense, rays of light are bent downward significantly as they leave the surface. At a certain point, the rays leaving in a horizontal direction "fall back." The star continues to collapse, the gravitational field becomes more powerful still, and light rays "fall back" at ever-increasing angles. When the radius of the star gets so small that the gravitational field strength at the surface reaches the critical threshold, only those photons traveling straight up from the surface manage to get away. The dying star has reached its event horizon! But things do not stop there. The collapsing star, or *collapsar*, keeps shrinking within the event horizon. Then, in simplistic theory, all photons emitted from the surface of the star are trapped in the black hole.

The idea of a black hole is not new. As long as there has been a particle theory of light, imaginative scientists have theorized that black holes can exist. When Albert Einstein revolutionized physics with his special and general theories of relativity in the early 1900s, new evidence arose for the existence of black holes. But nobody had ever seen an object through a telescope that truly fit the description. That was no surprise. The nature of black holes, assuming they exist, is such that they ought to be invisible at all electromagnetic wavelengths.

Critics of the black-hole theory said at first that the whole business was nonsense and the study of such things a waste of time. How, they asked, could you ever prove the existence of something that could never be seen? The answer to that question came much later, when indirect empirical evidence was found for the existence of black holes. Recently, most astronomers have come to believe that black holes do exist. In fact, the centers of many galaxies, including our own, are

thought to be undergoing gravitational collapse, harboring gigantic black holes made up of millions or even billions (thousand-millions) of stars.

THE SCHWARZCHILD RADIUS

A simple single-variable formula, based only on mass and two known constants, allows us to calculate the radius to which any spherical object would have to contract in order for the minimum escape speed to reach the speed of light in free space. In a simplistic sense, a sphere surrounding a point mass and having this radius represents an event horizon. This theoretical radius is called the *Schwarzchild radius*, named after the astrophysicist Karl Schwarzchild who first came up with the formula in 1916. He based the derivation of his formula on the supposition that if an object got small enough, the energy of photons emitted from the surface of that object would not be sufficient to propel them out of its gravitational field. Sometimes the Schwarzchild radius is called the *gravitational radius*.

Suppose an object has a mass m (in kilograms). Its Schwarzchild radius r (in meters) depends directly on the mass. The greater the mass, the greater the Schwarzchild radius. The formula is remarkably simple:

$$r = 2Gm / c^2$$

where c^2 is the speed of light squared (approximately 8.9875×10^{16} m²/s²) and G is the gravitational constant. In this formula, the value of the coefficient 2 can be considered exact.

Go ahead and perform some calculations with this formula. Do you want to figure out the Schwarzchild radius of your body? Remember, you must use your mass in kilograms to get the answer. You'll come up with a radius of submicroscopic proportions. The earth would have to be compressed to a radius of about 9 mm (3/8 in)—the size of a large marble—in order to fall within its Schwarzchild radius. The sun would have to collapse to a radius of approximately 2.8 km (1.7 mi). It is widely believed that a dying star must have three or more *solar masses* (that is, three or more times the mass of our sun) in order to collapse to the Schwarzchild radius.

The sequence of events that take place after the Schwarzchild radius has been reached depends on the point of view of the observer. Someone watching from the outside would see a collapsing star slow down in its rate of contraction as it approached the Schwarzchild radius, never quite reaching it. That is because extreme time dilation would occur as a result of the intense gravitation. At the

Schwarzchild radius, the gravitational time dilation function "blows up to infinity." If an observer could ride on the surface of a collapsing star and survive, no one knows what she would witness. But this is academic because the density of a collapsing star would produce gravitation intense enough to crush any passenger to death before the Schwarzchild radius was reached.

Interestingly, if a collapsing object or group of objects has enough mass, the density at the Schwarzchild radius is survivable, as we will shortly see. Perhaps a person falling within the event horizon of a collapsing galaxy would live to emerge in a different universe, closed off by an infinite space-time gulf from the universe she just left. But no one knows for sure.

NON-EUCLIDEAN SPACE

We ordinarily think of space as having three dimensions such as height, width, and depth. But according to Einstein, *three-space* is curved with respect to a fourth spatial dimension that we cannot see. We expect that rays of light in space should obey the rules of *Euclidean geometry*, which you learned in middle school or high school. One of the fundamental rules of light behavior, according to Newtonian physics, dictates that rays of light always travel in straight lines. Such a continuum is called *Euclidean space*. But scientists have seen that light beams don't always travel in straight lines! In order to explain this and certain other phenomena that astronomers have observed, we must allow for *non-Euclidean space.*

According to Einstein's theory of general relativity, space is curved or "warped" by gravitation. The extent of this curvature is insignificant at the earth's surface. It takes an enormously powerful gravitational field to cause enough curvature of space for its effects to be observed as an apparent bending of light rays. But the effect has been measured in starlight passing close to the sun during total solar eclipses, when the sun's disk is darkened enough so stars almost behind it can be seen through telescopes. The light from certain distant *quasi-stellar sources* (*quasars*), when passing near closer objects having intense gravitational fields, produces multiple images as the rays are bent.

The "warping" of space in the vicinity of a strong gravitational field has been likened to the stretching of a rubber membrane when a mass is placed on it. Suppose a thin rubber sheet is placed horizontally in a room, attached to the walls midway between the floor and the ceiling, and stretched until it is flat. Then a small, dense object such as a ball of iron is dropped in the center of the sheet. The sheet will become stretched into a funnel-like shape. The curvature of the sheet will be greatest near the ball, and will diminish with increasing distance from the

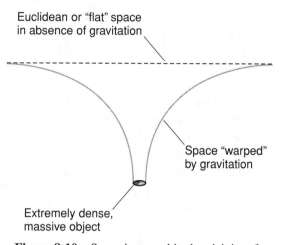

Figure 8-10   Space is curved in the vicinity of an object having an intense gravitational field.

ball. This is how Einstein believed that space could be distorted by an intense gravitational field (Fig. 8-10). But instead of two dimensions, the "rubber sheet" of space has three dimensions.

Any vessel or object traveling through "warped" space cannot travel in a true straight line because there are no straight paths within such a continuum. There exist only "shortest possible paths" called *geodesics* between arbitrary pairs of points. According to the theory of general relativity, real space is non-Euclidean everywhere, because there are gravitational fields everywhere. In some regions of space, gravitation is weak and in other places it is strong, but there is no place in the universe that entirely escapes this phenomenon.

GRAVITATIONAL WAVES

There are plenty of stars that are more than three times the mass of the sun. Astronomers agree that many of these stars have long since blown up, gone through their *white dwarf* phases, and presumably also gone through the collapsing process. This suggests that black holes make up a significant proportion of the matter in the universe.

As long as we can't travel to them or even see them, we have no good way to verify from direct experience that black holes exist. But fortunately, real black holes are not as black as theory implies. According to the research of the well-known cosmologist Stephen Hawking, black holes gradually lose their mass by emitting energy. This need not necessarily all be in the form of electromagnetic radiation such as visible light. Energy can also be lost as *gravitational waves*.

The idea of a gravitational wave was first made plausible when Einstein developed his general theory of relativity. Figure 8-11 is a simplified rendition of a gravitational wave as it leaves a black hole and travels through the *space-time continuum*. Just as a pebble, when dropped into a still pond, produces concentric, expanding, circular ripples on the two-dimensional surface of the water, so a black hole might produce concentric, expanding, spherical ripples in the three-dimensional continuum of space. Ripples would also occur in time! *Gravitational-*

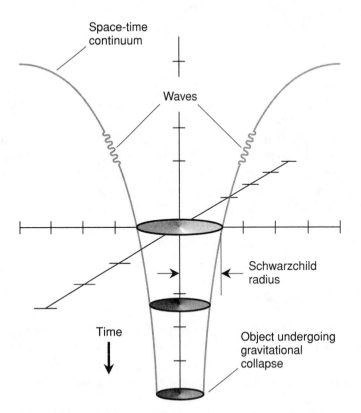

Figure 8-11　When an object undergoes gravitational collapse, it might emit gravitational waves.

wave detectors have been built in an effort to detect ripples in space and time as they pass. Because they involve the very fabric of the cosmos, such waves should be able to penetrate anything with no difficulty whatsoever. Therefore, a gravitational disturbance coming from the nadir (straight down as you see it while standing upright), and passing through the entire mass of our planet, ought to be as easily detected as one coming from the zenith (straight overhead).

PROBLEM 8-8

What is the magnitude of the gravitational force on a 1.00-kg object at the event horizon of a collapsed star with a mass of 7.60×10^{30} kg? Express the answer to three significant figures.

SOLUTION 8-8

First, we must calculate the Schwarzchild radius r, because that is the distance from the singularity at which the event horizon exists. Using the formula from above, let G be the gravitational constant, m be the mass of the collapsed star, and c^2 be the speed of light squared. Then:

$$
\begin{aligned}
r &= 2Gm \,/\, c^2 \\
&= 2.00 \times 6.67 \times 10^{-11} \times 7.60 \times 10^{30} / (8.9875 \times 10^{16}) \\
&= (1.01384 \times 10^{21}) / (8.9875 \times 10^{16}) \\
&= 1.13 \times 10^4 \text{ m}
\end{aligned}
$$

Next, we use the formula noted earlier in this chapter for the gravitational force F between two point sources as a function of their masses m_1 and m_2 and their separation s. Let m_1 be the mass of the collapsed star, m_2 be the mass of our small object, and s be the Schwarzchild radius, representing the distance between the singularity and our object. Then:

$$
\begin{aligned}
F &= Gm_1 m_2 \,/\, s^2 \\
&= 6.67 \times 10^{-11} \times 7.60 \times 10^{30} \times 1.00 / (1.13 \times 10^4)^2 \\
&= (5.0692 \times 10^{20}) / (1.2769 \times 10^8) \\
&= 3.97 \times 10^{12} \text{ N}
\end{aligned}
$$

That is approximately 4×10^{11} times the gravitational force that the same object would experience at the surface of the earth!

PROBLEM 8-9

Suppose that an arbitrary object of mass m (in kilograms) collapses to its Schwarzchild radius. Derive a formula for the density D of the resulting object (in kilograms per meter cubed) as a function of m. Assume that, when the object reaches its Schwarzchild radius, it is a perfect sphere. Express final coefficients, if any, to three significant figures.

SOLUTION 8-9

In order to determine this, first recall the formula for the volume V of a sphere in terms of its radius r:

$$V = 4\pi r^3 / 3$$

where V is in meters cubed and r is in meters. If such an object has mass m (in kilograms), then its density D (in kilograms per meter cubed) is:

$$D = m / V$$
$$= 3m / 4\pi r^3$$

Recall the formula for the Schwarzchild radius r in terms of mass m:

$$r = 2Gm / c^2$$

Substituting $(2Gm / c^2)$ for r in the previous formula for density, we obtain:

$$D = 3m / 4\pi (2Gm / c^2)^3$$
$$= 3c^6 / (32\pi G^3 m^2)$$

The values of G, c, and π are known constants. Approximately, we have:

$$G = 6.67 \times 10^{-11}$$
$$c = 2.9979 \times 10^8$$
$$\pi = 3.1416$$

Plugging in these values for the constants in the above equation for density gives us:

$$D = 3 \times (2.9979 \times 10^8)^6 / [32 \times 3.1416 \times (6.67 \times 10^{-11})^3 \times m^2]$$
$$= (2.1778 \times 10^{51}) / (2.9832 \times 10^{-29} \times m^2)$$
$$= (7.30 \times 10^{79}) / m^2$$

This means that the density of an object at its Schwarzchild radius varies inversely with the square of its mass. We should not be surprised at the huge size of the coefficient when we remember that the earth at its Schwarzchild radius would be smaller than a golf ball and yet contain a mass of six trillion-trillion kilograms! Such an object would be more dense than anyone can imagine. But as objects or collections of objects get more massive, their density at the Schwarzchild radius decreases. For huge masses, the *Schwarzchild density* can be quite manageable, as we shall now see.

PROBLEM 8-10

The mass of our sun, a typical star, is approximately 2.0×10^{30} kg and its radius is close to 7.0×10^8 m. The mass of a medium-sized spiral galaxy is on the order of 2.0×10^{11} times the mass of the sun. Suppose such a galaxy collapses to its Schwarzchild radius. How dense will the resulting object be, in kilograms per meter cubed? How does this compare with the average density of the sun, and therefore the average density of a typical star? Express the answers to two significant figures.

SOLUTION 8-10

According to the above data, the mass m_g of a medium-sized spiral galaxy is approximately:

$$m_g = 2.0 \times 10^{30} \times 2.0 \times 10^{11}$$
$$= 4.0 \times 10^{41} \text{ kg}$$

Using the formula for density at the Schwarzchild radius in terms of mass, we obtain:

$$D = (7.30 \times 10^{79}) / m_g^2$$
$$= (7.30 \times 10^{79}) / (4.0 \times 10^{41})^2$$
$$= (7.30 \times 10^{79}) / (1.6 \times 10^{83})$$
$$= 4.6 \times 10^{-4} \text{ kg/m}^3$$

This is far less than the density of the earth's atmosphere at the surface! Now let's calculate the average density D_s of the sun. Its radius is 7.0×10^8 m. That means the volume V_s of the sun is approximately:

$$V_s = 4.0 \times 3.1416 \times (7.0 \times 10^8)^3 / 3.0$$
$$= 4.0 \times 3.1416 \times 3.43 \times 10^{26} / 3.0$$
$$= 1.4 \times 10^{27} \text{ m}^3$$

The mass m_s of the sun is 2.0×10^{30} kg. Its approximate density can therefore be calculated as follows:

$$D_s = m_s / V_s$$
$$= (2.0 \times 10^{30}) / (1.4 \times 10^{27})$$
$$= 1.4 \times 10^3 \text{ kg/m}^3$$

This is close to the density of liquid water (1.0×10^3 kg/m^3), and roughly 3×10^6 times as dense as our galaxy would be if it were to contract to its Schwarzchild radius. This suggests that a galaxy undergoing gravitational collapse can reach and pass its Schwarzchild radius while its stars retain their individual identities. It also tempts us to imagine that a human astronaut could take a journey into a black hole of such mass and survive, at least for a little while.

Quiz

This is an "open book" quiz. You may refer to the text in this chapter. A good score is 8 correct. Answers are in the back of the book.

1. Neglecting the gravitational effects of the sun and the moon, and assuming the earth is a perfect sphere with uniform density throughout its interior, the minimum escape speed for an object starting at an initial altitude h above the surface
   (a) increases as h increases.
   (b) does not change as h increases.
   (c) decreases as h increases.
   (d) depends on the mass of the object.

2. Imagine two satellites X and Y revolving around the earth in perfectly circular orbits. Suppose the radius of the orbit of Y is exactly twice the radius of the orbit of X. If the tangential orbital speed of satellite X is 2.20×10^4 km/h, what is the tangential orbital speed of satellite Y? Neglect possible gravitational effects of the moon and the sun.
   (a) More information is needed to answer this.
   (b) 7.78×10^3 km/h.
   (c) 1.10×10^4 km/h.
   (d) 1.56×10^4 km/h.

3. Imagine an object at rest on a planet where the acceleration of gravity is 7.73 m/s². The weight of the object is found to be 480 N. What is its mass in pounds?
   (a) 28.2 lb.
   (b) 62.1 lb.
   (c) 137 lb.
   (d) In order to determine this, we must know the mass and the radius of the planet.

4. The acceleration of gravity at the Schwarzchild radius surrounding a singularity having 2.0×10^{11} solar masses would be approximately
   (a) 76 m/s².
   (b) 5.9×10^{14} m/s².
   (c) zero.
   (d) "infinity."

5. From the formulas given in this chapter, we can determine the theoretical tangential orbital speed of an object in a perfectly circular orbit around a much more massive space-time singularity when the orbital radius is equal to the Schwarzchild radius. This speed is independent of the mass of the central body. What is it?

   (a) The speed of light in free space.

   (b) The speed of light in free space, divided by the square root of 2.

   (c) Half the speed of light in free space.

   (d) One-quarter of the speed of light in free space.

6. Neglecting the gravitational effects of the sun and the moon, and assuming the earth is a perfect sphere with uniform density throughout its interior, an object in an elongated, elliptical orbit around the earth would

   (a) continuously gain potential energy, but at a variable rate.

   (b) continuously lose potential energy, but at a variable rate.

   (c) alternately gain and lose potential energy.

   (d) always have the same potential energy.

7. Imagine a solid object having constant mass that starts out at the earth's surface and rises straight up. Neglecting the effects of the sun and the moon, how can we describe the way in which the gravitational potential energy changes with altitude?

   (a) The object loses potential energy at a constant rate with increasing altitude.

   (b) The object loses potential energy at a decreasing rate with increasing altitude.

   (c) The object loses potential energy at an increasing rate with increasing altitude.

   (d) The object gains potential energy as the altitude increases.

8. Suppose the acceleration of gravity is 6.400 m/s^2 on the surface of a planet having a radius of 4000 km. For mathematical convenience, consider all the planet's mass to be concentrated at its center. Now imagine an object in a perfectly circular orbit 2000 km above the surface. What is the acceleration of gravity on this object?

   (a) 4.267 m/s^2.

   (b) 2.844 m/s^2.

   (c) Zero, because the object is "weightless."

   (d) In order to determine this, we must know the mass of the planet.

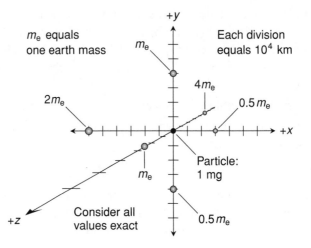

Figure 8-12   Illustration for Quiz Question 9.

9. Refer to Fig. 8-12. Given this information, the direction of the net gravitational force vector imposed on the 1-mg particle by the six massive objects in its vicinity is:

(a) in the direction of the positive z axis.

(b) in the direction of the positive y axis.

(c) in the direction of the positive x axis.

(d) in no direction, because there is no net gravitational force on the 1-mg particle.

10. Imagine a hypothetical object whose mass can be varied at will. As the mass of this object changes, its Schwarzchild radius varies

(a) directly in proportion to the square of its mass.

(b) directly in proportion to its mass.

(c) inversely in proportion to its mass.

(d) inversely in proportion to the square of its mass.

CHAPTER 9

Atomic Physics and Radiant Energy

In this chapter we'll explore the ongoing mysteries concerning what matter and energy actually are. We'll also learn how matter and energy are different and at the same time similar. Are they both manifestations of the same thing?

The Nature of Matter

An *atom* is the smallest sample of a given *chemical element* that can exist while allowing that element to retain its identity. The slightest change in the internal structure of an atom can make a huge difference in its behavior. Let's review the general structure of atoms. Then we'll look at some phenomena that support the *matter-energy duality*.

THE NUCLEUS

The *nucleus* is the part of an atom that gives a chemical element its identity. An atomic nucleus is made up of two types of particles, the *neutron* and the *proton*. A neutron, which carries no electrical charge, has slightly more mass than a proton, which carries a small positive electric charge. Both of these particles are far more dense than any sample of matter you see in the everyday world. A teaspoonful of either of these particles, packed tightly together, would mass thousands of kilograms.

The mass of a neutron at rest (m_n) and the mass of a proton at rest (m_p) are specified by NIST as follows, accurate to six significant figures:

$$m_n = 1.67493 \times 10^{-27} \text{ kg}$$
$$m_p = 1.67262 \times 10^{-27} \text{ kg}$$

The number of protons in an elemental nucleus is called the *atomic number*. The element with one proton in its nucleus is *hydrogen*; the element with two protons in the nucleus is *helium*. Both of these are gases at standard temperature and pressure. If there are three protons, we have *lithium*, a light metal that combines easily with many other elements. The element with four protons is *beryllium*, also a light metal. The atomic number uniquely determines the identity of an element. If a nucleus has one proton, we can be certain that it's a hydrogen nucleus; if a nucleus contains two protons, we know it's a helium nucleus, and so on.

In general, as the number of protons in an atomic nucleus increases, the number of neutrons also increases, although there are some exceptions. Elements with high atomic numbers such as *uranium* (atomic number 92) are therefore more dense, when found in nature, than elements with low atomic numbers such as *carbon* (atomic number 6).

ISOTOPES

For any particular chemical element, the number of protons is always the same, but the number of neutrons can vary. Differing numbers of neutrons result in different *isotopes* for a given element. All elements have at least two isotopes, and some have dozens.

Every element has one isotope that is most often found in nature. A change in the number of neutrons in the nucleus of an element results in a change in the density of the substance as it exists in the real world. Certain other characteristics can change, as well. Some isotopes stay the same over long periods of time; these are

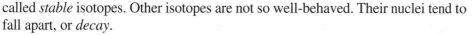

called *stable* isotopes. Other isotopes are not so well-behaved. Their nuclei tend to fall apart, or *decay*.

When atomic nuclei decay spontaneously, they are said to be *unstable*. The decay process is always attended by the emission of radiant energy and/or subatomic particles. These emissions are known as *radioactivity*. Among laypeople, perhaps the most well-known example of a natural radioactive element is uranium. All of its known isotopes are unstable, but some decay more rapidly than others.

ATOMIC WEIGHT

The *atomic weight* of an element is approximately (or exactly, in the case of the most common natural isotope of carbon) equal to the sum of the number of protons and the number of neutrons in the nucleus. As you learned in Chapter 1, one atomic mass unit (amu or u) is defined as precisely 1/12 of the mass of a nucleus of carbon-12 or C-12, which has six protons and six neutrons.

Some atoms of carbon have eight neutrons in the nucleus rather than six. This type of carbon atom has an atomic weight of 14, and is known as carbon-14 or C-14. It is unstable, although it decays at a slow rate. It is used by scientists to measure how old things are in geological terms.

Uranium can exist in the form of U-234 (atomic weight 234) or U-235 (atomic weight 235) as well as the most common isotope, U-238 (atomic weight 238). In a *uranium-fission power plant*, U-235 is the isotope of choice because it has properties that make it possible to induce controlled decay, producing a steady supply of energy that can be harnessed.

ELECTRONS AND SHELLS

An *electron* is a subatomic particle with negative electric charge and only a small fraction of the mass of a neutron or proton. An electron has the same charge quantity as a proton, but with negative polarity rather than positive polarity. The charge quantity on a single electron or proton is the smallest possible amount of electric charge—the elementary charge unit (ecu) as you learned in Chapter 5.

The NIST specifies the mass of an electron at rest (m_e), accurate to six significant figures, as follows:

$$m_e = 9.10938 \times 10^{-31} \text{ kg}$$

According to an early theory concerning the nature of matter, electrons were believed to be embedded in atoms, like raisins in a fruitcake. In fact, this concept gave rise to the term *raisin-cake model*. Later, physicists and chemists learned that all atoms in the ordinary world consist almost entirely of space between nuclei and electrons. Electrons were envisioned as orbiting the nucleus, the atom resembling a miniature "solar system" with the nucleus as the "sun" and the electrons as the "planets" (Fig. 9-1). Atomic nuclei carry a net positive charge, while electrons carry a negative charge. Particles having opposite electrical charges attract each other. So instead of gravitation keeping electrons in orbit the way the sun holds the planets, electrical attraction is at work, keeping electrons in orbit around a single nucleus—usually.

The "solar system" model is an oversimplification. In the modern view, electrons are envisioned as "swarming" around atomic nuclei so fast, and in such a complicated way, that it is impossible to pinpoint the location of any single electron at any point in time. All that can be said is that, at any moment, an electron is just as likely to be located inside a certain sphere as outside it. A sphere thus defined is called an *electron shell*. In simplistic terms, the center of an electron shell normally corresponds to the location of the atomic nucleus.

The larger the radius of an electron shell, the more energy the electron has. Electron shells are assigned so-called *principal quantum numbers*. The lowest principal quantum number is $n = 1$, which represents the smallest possible electron shell. Progressively larger principal quantum numbers $n = 2$, $n = 3$, $n = 4$, ... represent progressively larger shells. Therefore, as the principal quantum number of an electron increases, so does its energy, if all other factors are held constant.

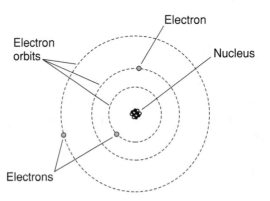

Figure 9-1 An early model of the atom, developed around the year 1900. The electrons were thought to orbit the nucleus like planets orbit the sun.

IONS

Under some conditions, electrons are stripped "out of orbit." Then they roam among nuclei or travel freely through space. This produces atoms in which the number of electrons differs from the number of protons. A shortage of electrons gives an atom a net positive charge; an excess of electrons gives an atom a net negative charge. The element's identity remains the same, however, regardless of how great the excess or shortage of electrons might be.

Sometimes, atomic nuclei get stripped of all their electrons. A common example is a nucleus of helium-4 (He-4) which has two protons and two neutrons. When this nucleus travels at high speed, it is called an *alpha* (α) *particle*, and constitutes a form of radioactivity.

An electrically charged atom is called an *ion*. When a substance contains many ions, the material is said to be *ionized*. When some of the atoms in a sample of matter become ionized, the process responsible for this change is known as *ionization*. Ionized materials generally conduct electricity well, even if the substance is normally not a good conductor.

An element can be both an ion and an isotope different from the usual isotope. For example, an atom of carbon might have eight neutrons rather than the usual six, thus being the isotope C-14, and it might have been stripped of an electron, giving it a positive unit electric charge and making it an ion.

THE BOHR MODEL

Electron shells always have specific radii. They don't occur at random. The discovery of this phenomenon is credited to Niels Bohr, who noticed it in the early part of the twentieth century. His concept of the atom became known as the *Bohr model* or the *Bohr atom*.

In the Bohr model, electrons can move from smaller shells to larger ones (Fig. 9-2) or vice-versa. When an electron gains energy by absorbing an "energy packet" called a *photon* having just the right *electromagnetic* (EM) *wave frequency*, that electron moves from a shell with a certain radius to a shell with a larger radius. When an electron moves from a larger shell to a smaller one, it loses energy and emits a photon having a specific EM wave frequency. These frequencies are predictable because the radii of the shells are predictable. For a given chemical element, the set of EM waves thus produced comprise a "signature" called an *absorption spectrum* or *emission spectrum*. These spectra can be used to identify the existence of various elements in distant objects such as stars, nebulae, and galaxies.

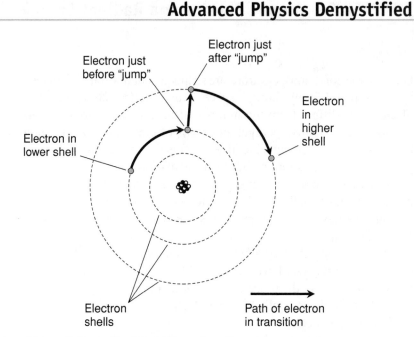

Figure 9-2 A simplified illustration of an electron gaining energy within an atom.

In the Bohr model of the hydrogen atom, the radius of the lowest-energy electron shell is a constant that is sometimes used as a standard unit of length at the atomic scale. This constant is called the *Bohr radius* (a_o) and is given by NIST to six significant figures as follows:

$$a_o = 5.29177 \times 10^{-11} \text{ m}$$

THE EINSTEIN FORMULA

Nearly everyone has heard or read the equation derived by Albert Einstein to express the relationship among mass m, energy E, and the speed of light c in free space:

$$E = mc^2$$

where E is in joules, m is in kilograms, and c is in meters per second. Conversely:

$$m = E / c^2$$

These equations quantify various phenomena such as reactions that take place in the sun, in atomic power plants, and in hydrogen bombs.

If you could convert a particle of matter entirely to energy, the amount of energy could be calculated using the so-called *Einstein formula*. But this sort of conversion—a free particle merely vanishing in a "flash of light" or a "burst of radiation"—rarely takes place. Instead, energy is produced or absorbed, and shows up as a discrepancy between the masses of particles before and after a *nuclear reaction*.

NUCLEAR FORCE AND BINDING ENERGY

Particles having opposite electric charges attract, while particles having like charges repel. The strength of this electric force is directly proportional to the product of the charge quantities, and inversely proportional to the square of the distance between the charge centers. In an atomic nucleus having more than one proton, multiple positively charged particles are extremely close to each other. According to the laws of electrostatics, they ought to fly apart because of electrical repulsion. But they are held together by an attractive force called the *strong nuclear force* that's far more powerful at close range than the electric force. This same force keeps neutrons and protons in an atomic nucleus from spontaneously drifting apart.

Consider a *deuteron*, which consists of one neutron and one proton held together by the strong nuclear force. A deuteron is a nucleus of *deuterium*, an isotope of hydrogen with an atomic weight approximately equal to 2. It seems reasonable to suppose that the rest mass m_d of a deuteron ought to be equal to the sum of the rest mass m_n of a neutron and the rest mass m_p of a proton. But in fact it is a little less. We can express the relationship as follows:

$$m_d = (m_n + m_p) - \Delta m$$

where Δm is a "fudge factor" called the *mass defect*.

The sum of the rest masses of a single proton and a single neutron, considered separately, is:

$$m_n + m_p = 1.67493 \times 10^{-27} + 1.67262 \times 10^{-27}$$
$$= 3.34755 \times 10^{-27} \text{ kg}$$

The rest mass of a deuteron has been measured and found to be as follows, accurate to six significant figures according to NIST:

$$m_d = 3.34358 \times 10^{-27} \text{ kg}$$

You can perform the subtraction and verify that m_d is 3.97×10^{-30} kg less than $m_n + m_p$. This difference constitutes the mass defect Δm.

Now imagine that we force a neutron and a proton together to form a deuteron. In this process, 3.97×10^{-30} kg of mass "gets lost." We can't dismiss this discrepancy the way we use "fudge factors" in casual life, nor can we claim that a little bit of matter disappears into some cosmic underworld. The matter turns into energy known as *binding energy*, which is liberated when the neutron and proton combine or *fuse* to form a single deuteron.

If we let E_b represent the binding energy liberated when a neutron and a proton fuse to form a deuteron, and if we consider the speed of light in free space to be 3.00×10^8 m/s, we have:

$$
\begin{aligned}
E_b &= (\Delta m)\ c^2 \\
&= 3.97 \times 10^{-30} \times (3.00 \times 10^8)^2 \\
&= 3.97 \times 10^{-30} \times 9.00 \times 10^{16} \\
&= 3.57 \times 10^{-13}\ \text{J}
\end{aligned}
$$

If we want to break a deuteron apart into a separate neutron and proton, we must supply this same amount of energy.

The energy required to break a nucleus apart gives scientists a good idea of how powerful the strong nuclear force truly is. In order to obtain energy levels sufficient to break up a nucleus, subatomic particles can be propelled into that nucleus at high speed. Physicists do this in gigantic systems known as *particle accelerators*. The energy in a stream of protons or alpha particles, traveling at an appreciable fraction of the speed of light, is considerable.

Binding energy can manifest itself as visible light, *infrared* (IR), *radio waves*, *ultraviolet* (UV), *X rays*, or *gamma rays*, all of which are forms of *electromagnetic* (EM) radiation. Binding energy can also show up, or be absorbed, in the form of high-speed subatomic particles. Exotic particles such as the *positron* (or *anti-electron*), the *neutrino*, and the *anti-neutrino* can be part of such reactions.

THE ELECTRON-VOLT

At the atomic level, the joule is a huge unit of energy. Nuclear physicists use a smaller unit known as the *electron-volt* (eV) and power-of-10 multiples of it. The electron-volt is more reasonable in size for quantifying energy in atomic reactions, and it is more meaningful too. One electron-volt is the amount of energy acquired or gained by an electron when it passes through an electric potential difference of one volt (1 V) in free space. Expressed to six significant figures:

$$1\ \text{eV} = 1.60218 \times 10^{-19}\ \text{J}$$

The inverse relation is:

$$1 \text{ J} = 6.24151 \times 10^{18} \text{ eV}$$

Do you notice something familiar about these numbers? The ratio of the electron-volt to the joule is the same as the ratio of the elementary charge unit to the coulomb. This fact can help you to remember the conversion factors. As an equation in units only, we can write:

$$\text{ev} / \text{J} = \text{ecu} / \text{C}$$

which reads, "Electron-volts are to joules, as elementary charge units are to coulombs."

In some applications, power-of-10 multiples of the electron-volt are specified. You will occasionally read or hear about the *kiloelectron-volt* (keV), the *megaelectron-volt* (MeV), or the *gigaelectron-volt* (GeV). The relationships are:

$$1 \text{ keV} = 1000 \text{ eV}$$
$$1 \text{ MeV} = 1000 \text{ keV} = 10^6 \text{ eV}$$
$$1 \text{ GeV} = 1000 \text{ MeV} = 10^9 \text{ eV}$$

PROBLEM 9-1

How much energy, in megaelectron-volts, is emitted when a proton and a neutron fuse to form a deuteron? Express the answer to three significant figures.

SOLUTION 9-1

We determined earlier that the binding energy liberated when a neutron and a proton fuse to form a deuteron is:

$$E_b = 3.57 \times 10^{-13} \text{ J}$$

If we want to convert this to electron-volts, we must multiply by the number of electron-volts in a joule. Then, to get megaelectron-volts, we divide the value in electron-volts by 10^6 (or multiply by 10^{-6}). Calculating:

$$E_b = 3.57 \times 10^{-13} \times 6.24151 \times 10^{18}$$
$$= 3.57 \times 6.24151 \times 10^{-13} \times 10^{18}$$
$$= 2.23 \times 10^6 \text{ eV}$$
$$= 2.23 \text{ MeV}$$

PROBLEM 9-2

Suppose that you could figure out a way to fuse an electron and a proton and end up with a neutron. Would this hypothetical process require energy or liberate energy? How much energy? Express the answer to three significant figures, in joules and also in kiloelectron-volts. Consider $c = 3.00 \times 10^8$ m/s.

SOLUTION 9-2

First, let's add the rest mass m_p of a proton to the rest mass m_e of an electron:

$$m_p + m_e = 1.67262 \times 10^{-27} + 9.10938 \times 10^{-31}$$
$$= 1.673530938 \times 10^{-27} \text{ kg}$$

The rest mass m_n of a neutron is given by NIST as:

$$m_n = 1.67493 \times 10^{-27} \text{ kg}$$

This is greater than the sum of the rest masses of a proton and an electron. In order to combine them and get a neutron, we would have to supply energy equivalent to the mass difference Δm. By subtraction:

$$\Delta m = m_n - (m_p + m_e)$$
$$= 1.67493 \times 10^{-27} - 1.673530938 \times 10^{-27}$$
$$= (1.67493 - 1.673530938) \times 10^{-27}$$
$$= 1.399062 \times 10^{-30} \text{ kg}$$

Using the Einstein equation to convert this mass to its equivalent energy in joules (let's call it E_{pen}), we obtain:

$$E_{pen} = (\Delta m) \, c^2$$
$$= 1.399062 \times 10^{-30} \times (3.00 \times 10^8)^2$$
$$= 1.399062 \times 10^{-30} \times 9.00 \times 10^{16}$$
$$= 1.26 \times 10^{-13} \text{ J}$$

In order to convert this to kiloelectron-volts, we must multiply the above value by 6.24151×10^{18}, and then divide the result by 10^3 (or multiply by 10^{-3}). Calculating:

$$E_{pen} = 1.26 \times 10^{-13} \times 6.24151 \times 10^{18} \times 10^{-3}$$
$$= 1.26 \times 6.24151 \times 10^{-13} \times 10^{18} \times 10^{-3}$$
$$= 786 \text{ keV}$$

The Nature of Radiant Energy

Radiant energy exhibits wavelike properties such as *wavelength*, *interference* (phase cancellation and reinforcement), and *diffraction* (the ability to "bend" around sharp corners). This behavior led nineteenth-century scientists to develop the *wave theory of light*, which extends to all forms of EM radiation. But radiant energy also propagates as if it were a stream of high-speed particles. Isaac Newton thought of that idea, and Albert Einstein refined it.

WAVE "PACKETS"

Even in the 1800s, physicists knew that radiant energy could ionize the atoms of metals, changing their electrical conductance. This was believed to result from EM waves imparting energy to electrons, causing them to move to larger shells within atoms or to escape from the nuclei and become free. A *free electron* can move easily from one atom to another when a voltage appears, or is applied, between different points in a sample of matter. A material that has a certain electrical conductance in total darkness may have improved conductance when visible, IR, UV, or other EM rays shine on its surface. This is called the *photoelectric effect*. It occurs because of ionization.

In the early twentieth century, Einstein concluded that the wave theory of light could not adequately account for the photoelectric effect. According to the wave theory, the photoelectric effect should occur in any metal if the source of the radiation is intense enough. Experimentation revealed that things don't work that way. Instead, the occurrence (or absence) of the photoelectric effect correlates more closely with the frequency of EM radiation than with its intensity. A metal that shows pronounced photoelectric behavior for visible light at moderate intensity might not exhibit it when bombarded with radio microwaves, no matter how powerful. Conversely, materials that do not ionize under bright visible light might readily ionize when exposed to relatively weak X rays.

In order to explain the photoelectric effect fully, Einstein theorized that EM radiation consists of discrete "packets of wavelets." The term *photon* was coined for these "packets" because they were originally conceived in the *particle theory of light*. (The prefix *photo-* refers to visible light.) Today, the particle theory applies to all EM radiation, visible and invisible. The particle theory offers a plausible explanation of the anomalies of the photoelectric effect, and also allows for all the observed wavelike characteristics of radiant energy.

FREQUENCY, PERIOD, AND WAVELENGTH

Let's review the relationships among the properties of EM waves in free space. Besides frequency, the *period* and the *wavelength* are often specified, assuming a *propagation speed* equal to the speed of light c, which is approximately 3.00×10^8 m/s in free space. The period T (in seconds) of an EM wave is the reciprocal of the frequency f (in hertz). Mathematically:

$$f = 1 / T$$
$$T = 1 / f$$

The period of a wave is related to the wavelength λ (in meters) and the propagation speed c (in meters per second) as follows:

$$\lambda = cT$$
$$T = \lambda / c$$

This gives rise to other formulas:

$$\lambda = c / f$$
$$f = c / \lambda$$
$$c = f\lambda$$
$$c = \lambda / T$$

Of primary interest is the relationship between frequency and wavelength. In general, as the frequency goes up, the wavelength becomes shorter. You can use these formulas:

$$\lambda = 3.00 \times 10^8 / f$$
$$f = 3.00 \times 10^8 / \lambda$$

where λ is the wavelength in meters and f is the frequency in hertz. If you need greater accuracy, you can use the NIST value for the speed of light in free space, expressed to six significant figures as:

$$c = 2.99792 \times 10^8 \text{ m/s}$$

PHOTON "MASS" AND ENERGY

If we accept the notion that photons are particles, it is tempting to think they have mass. But the term *rest mass* is meaningless for photons because whenever we find one, it's moving. Photons are never at rest! The common consensus among physicists these days is that a photon is *massless*. That is not to say that it has a mass equal to zero. It means that we might as well forget about trying to give it a real-number mass at all.

When a particle with rest mass m moves at an extremely high speed v, then the mass in effect increases by a *relativistic factor*, becoming a larger mass m_{rel} according to this formula:

$$m_{\text{rel}} = m / (1 - v^2 / c^2)^{1/2}$$

When $v = c$, the equation "blows up":

$$
\begin{aligned}
m_{\text{rel}} &= m / (1 - v^2 / c^2)^{1/2} \\
&= m / (1 - c^2 / c^2)^{1/2} \\
&= m / (1 - 1)^{1/2} \\
&= m / 0^{1/2} \\
&= m / 0
\end{aligned}
$$

If we let the rest mass m_{ph} of the photon be equal to zero, then according to relativity theory, its mass m_{ph-rel} at the speed of light (which is always the speed of a photon in free space) is zero divided by zero:

$$m_{ph-rel} = m_{ph} / 0$$
$$= 0 / 0$$

That's mathematical nonsense! But if we let m_{ph} be anything larger than zero, then m_{ph-rel} becomes the quotient of some positive real number over zero. Again, nonsense.

While we can't say much about photon mass, we can determine the energy E_{ph} in a photon if we know the frequency f of the EM wave associated with it:

$$E_{ph} = hf$$

where E_{ph} is in joules, f is in hertz, and h is a tiny number known as *Planck's constant*. Expressed to six significant figures according to NIST:

$$h = 6.62607 \times 10^{-34} \text{ J} \cdot \text{s}$$

If we know the wavelength λ associated with a photon instead of the frequency, we can substitute in the above equation, obtaining:

$$E_{ph} = h \,(3.00 \times 10^8 / \lambda)$$
$$= 3.00 \times 10^8 \, h / \lambda$$

where E_{ph} is in joules and λ is in meters.

THE RYDBERG-RITZ FORMULA

Imagine a glass cylinder from which all the air has been pumped out. Suppose that a small amount of hydrogen gas is introduced, and then a DC electrical source is connected to electrodes at either end of the cylinder. This apparatus is called a *gas-discharge tube*. Such a tube can be constructed in a good high-school physics lab. If the DC source voltage E is high enough, the gas inside the tube ionizes and an electrical current I flows. As a result, a certain amount of power P is dissipated in the gas:

$$P = EI$$

where P is in watts, E is in volts, and I is in amperes.

One *watt* (1 W) of power is the equivalent of energy dissipated at the rate of one *joule per second* (1 J/s). The gas in the tube therefore absorbs energy as time passes. This causes electrons in the atoms to attain higher energy levels, and therefore to reside in larger shells, than would be the case if there were no voltage source connected to the electrodes. But this one-way process can't go on forever. As current flows through the ionized hydrogen gas, electrons not only rise to higher energy levels (larger shells) but also constantly fall back to lower energy levels (smaller shells), emitting photons at various EM wavelengths. Once an electron has fallen, it is ready to absorb some energy from the DC source again, rise again, and then fall again. After a short initial "warm-up time," the amount of energy absorbed by the gas from the DC source, causing electrons to rise from smaller shells to larger ones, balances the energy radiated as photons by the gas as electrons fall from larger shells back to smaller ones.

There are multiple electron shells in a hydrogen atom, so there are a great many different energy quantities that an electron can gain or lose as it moves from one shell to another. Because of this, energy is radiated from ionized hydrogen gas in the form of photons having numerous wavelengths λ as follows:

$$\lambda = 1 / [R_{H} (1/n_1{}^2 - 1/n_2{}^2)]$$

where λ is in meters, R_H is a constant called the *Rydberg constant*, n_1 is the principal quantum number of the smaller shell to which a particular electron descends, and n_2 is the principal quantum number of the larger shell from which that same electron has just fallen.

The above equation is known as the *Rydberg-Ritz formula*, named after Johannes Rydberg and Walter Ritz who first published it in 1888. The value of the Rydberg constant, given by NIST to six significant figures, is:

$$R_{H} = 1.09737 \times 10^7$$

This is often rounded to 1.10×10^7 and is defined in units of *per meter* (/m or m^{-1}). Some texts denote the Rydberg constant as R with an infinity-symbol subscript (R_∞). Once in a while you will see it symbolized simply as R, but you must be careful in that case not to confuse it with the universal gas constant.

THE BALMER FORMULA

The *Balmer formula* is a special case of the Rydberg-Ritz formula for $n_1 = 2$. This expresses a characteristic set of visible wavelengths that appear in the emission spectrum of ionized hydrogen gas:

$$\lambda = 1 / [R_{H} (1/4 - 1/n_2{}^2)]$$

where n_2 is an integer with a value of 3 or greater. Johann Balmer discovered this formula in 1885, three years before it was generalized by Rydberg and Ritz. The set of emission wavelengths it defines, which was observed for years before it was mathematically quantified, is known as the *Balmer series*.

Electrons are confined to shells, producing or absorbing energy at specific wavelengths when "falling" or "jumping" from one shell to another, because the angular momentum of any *bound electron* (an electron held captive by an atomic nucleus) must always be equal to the product of a positive integer, Planck's constant, and 2π. If L_e is the angular momentum of an electron, then:

$$L_e = n' \, (h / 2\pi)$$

where L_e is in kilogram meters squared per second (kg · m²/s) or joule-seconds (J · s), n' is an element of the set of positive integers $N = \{1, 2, 3, \ldots\}$, h is in joule-seconds, and π is the ratio of the circumference of a sphere to its diameter, the familiar constant approximately equal to 3.14159. The ratio of Planck's constant to 2π is called *h-bar* and is symbolized as a lowercase italic letter h with a horizontal bar through its stem ($\hbar$). This number, like Planck's constant itself, appears frequently in nuclear and quantum physics. To six significant figures:

$$\hbar = 1.05457 \times 10^{-34} \text{ J} \cdot \text{s}$$

Therefore:

$$L_e = n' \, \hbar$$

where L_e is in kilogram meters squared per second (kg · m²/s) or joule-seconds (J · s) and n' is an element of the set of positive integers $N = \{1, 2, 3, \ldots\}$.

DEBROGLIE WAVES

If the wavelength is short enough and if the radiation is intense enough, a burst of EM energy can "push" an electron into a higher orbit around a nucleus or "knock" the electron out of orbit altogether. According to the particle theory of EM radiation, this occurs because photons have momentum that they can impart to electrons or other particles. Even though it is difficult to define the mass of a photon, its momentum can be determined by experiment, and it is inversely proportional to the wavelength. The momentum p_{ph} of a photon is equal to Planck's constant h divided by the EM wavelength λ, as follows:

$$p_{ph} = h / \lambda$$

where p_{ph} is in kilogram meters per second, h is in joule-seconds, and λ is in meters. The momentum of a photon is directly proportional to the frequency. If we know the EM wave frequency f, then:

$$p_{ph} = h / (3.00 \times 10^8 / f)$$
$$= hf / (3.00 \times 10^8)$$

where p_{ph} is in kilogram meters per second, h is in joule-seconds, and f is in hertz.

If EM disturbances can be regarded as "packets of wavelets" having finite, non-zero momentum, it is reasonable to wonder if all moving particles have wave-like properties. Early in the twentieth century, the physicist Louis deBroglie (pronounced "LOO-ie de-BROY-lee") sought an answer to that question and decided that the answer must be "Yes." His theory, made public in 1923, became known as the *deBroglie hypothesis*. Since deBroglie's time, scientists have confirmed that we can apply the relation between momentum and wavelength, or the relation between momentum and frequency, to any moving particle. This gives us figures called the *deBroglie wavelength* and the *deBroglie frequency* for the particle:

$$\lambda = h \,/\, p$$
$$f = pv \,/\, h$$

Recall the formula for the linear momentum of a particle or object with rest mass m traveling at constant speed v in a straight line:

$$p = mv$$

where p is in kilogram meters per second, m is in kilograms, and v is in meters per second. Unlike photons, any particle of matter has a finite, non-zero mass. As long as v is a small fraction of the speed of light in free space, we can substitute in the above equations for deBroglie wavelength and frequency, obtaining:

$$\lambda = h \,/\, (mv)$$
$$f = mv^2 \,/\, h$$

where λ is in meters, f is in hertz, h is in joule-seconds, m is in kilograms, and v is in meters per second.

If the speed v of a moving particle is an appreciable fraction of the speed of light, then the rest mass m must be multiplied by a relativistic factor r_m, where:

$$r_m = 1 \,/\, (1 - v^2 \,/\, c^2)^{1/2}$$

Substituting in the above equations for wavelength and frequency in terms of rest mass and speed, we get:

$$\lambda = h \,[(1 - v^2 \,/\, c^2)^{1/2}] \,/\, (mv)$$
$$f = mv^2 \,/\, [h \,(1 - v^2 \,/\, c^2)^{1/2}]$$

where λ is in meters, f is in hertz, h is in joule-seconds, m is in kilograms, v is in meters per second, and c is approximately equal to 3.00×10^8 m/s.

These formulas can be theoretically applied to common objects as well as to subatomic particles. But figuring out the deBroglie wavelengths of moving objects in the everyday world is rarely done except for amusement. When we do this, we get ridiculously short wavelengths and high frequencies, far beyond the values for the most energetic and deadly known EM rays. (As an optional exercise, try calculating the deBroglie wavelength and frequency for a 5000-kg truck trundling down a straight highway at a constant speed of 25 m/s.) The notion that truck drivers routinely pilot "massive particles" at "deadly wavelengths" might give rise to some good science fiction, but it isn't of much use to the physicist.

PROBLEM 9-3

Visible light occurs at free-space EM wavelengths ranging from approximately 750 nanometers (nm) to 400 nm, where 1 nm = 10^{-9} m. What is this range in terms of frequency and period? Express the answers in hertz and in seconds, to three significant figures. Consider the speed of light in free space to be 3.00×10^8 m/s.

SOLUTION 9-3

First, we must convert the wavelengths to meters. Let λ_1 be the longest visible wavelength and let λ_2 be the shortest. This gives us $\lambda_1 = 7.50 \times 10^{-7}$ m and $\lambda_2 = 4.00 \times 10^{-7}$ m. Expressing these values as a range of frequencies where f_1 is lowest and f_2 is highest:

$$f_1 = c / \lambda_1$$
$$= (3.00 \times 10^8) / (7.50 \times 10^{-7})$$
$$= 4.00 \times 10^{14} \text{ Hz}$$

$$f_2 = c / \lambda_2$$
$$= (3.00 \times 10^8) / (4.00 \times 10^{-7})$$
$$= 7.50 \times 10^{14} \text{ Hz}$$

To determine the range of periods where T_1 is the longest and T_2 is the shortest, we can use the formulas for period in terms of frequency:

$$T_1 = 1 / f_1$$
$$= 1.00 / (4.00 \times 10^{14})$$
$$= 2.50 \times 10^{-15} \text{ s}$$

$$T_2 = 1 / f_2$$
$$= 1.00 / (7.50 \times 10^{14})$$
$$= 1.33 \times 10^{-15} \text{ s}$$

PROBLEM 9-4

What is the range of photon energy values for visible light? Use the frequencies f_1 and f_2 as derived in Solution 9-3. Express the answers in joules to three significant figures. Consider Planck's constant to be 6.63×10^{-34} J · s.

SOLUTION 9-4

Let $E_{\text{ph-1}}$ be the energy contained in a photon having frequency f_1, and let $E_{\text{ph-2}}$ be the energy contained in a photon having frequency f_2. Then:

$$E_{\text{ph-1}} = hf_1$$
$$= 6.63 \times 10^{-34} \times 4.00 \times 10^{14}$$
$$= 2.65 \times 10^{-19} \text{ J}$$

$$E_{\text{ph-2}} = hf_2$$
$$= 6.63 \times 10^{-34} \times 7.5 \times 10^{14}$$
$$= 4.97 \times 10^{-19} \text{ J}$$

PROBLEM 9-5

What is the range of photon momentum values for visible light? Use the wavelengths $\lambda_1 = 7.50 \times 10^{-7}$ m and $\lambda_2 = 4.00 \times 10^{-7}$ m as stated at the beginning of Solution 9-3. Express the answers in kilogram meters per second to three significant figures. Consider Planck's constant to be 6.63×10^{-34} J $\cdot$ s.

SOLUTION 9-5

Let $p_{\text{ph-1}}$ be the momentum of a photon having wavelength λ_1, and let $p_{\text{ph-2}}$ be the momentum of a photon having wavelength λ_2. Then:

$$p_{\text{ph-1}} = h / \lambda_1$$
$$= (6.63 \times 10^{-34}) / (7.50 \times 10^{-7})$$
$$= 8.84 \times 10^{-28} \text{ kg} \cdot \text{m/s}$$

$$p_{\text{ph-2}} = h / \lambda_2$$
$$= (6.63 \times 10^{-34}) / (4.00 \times 10^{-7})$$
$$= 1.66 \times 10^{-27} \text{ kg} \cdot \text{m/s}$$

In passing, note that $\lambda_1, f_1, E_{\text{ph-1}}$, and $p_{\text{ph-1}}$ in the preceding three problems represent parameters for visible red light, while $\lambda_1, f_2, E_{\text{ph-2}}$, and $p_{\text{ph-2}}$ represent parameters for visible violet light. In between these extremes as the wavelength decreases, we see the intermediate "rainbow colors": orange, yellow, green, blue, and indigo.

Nuclear Decay and Fission

Because radioactivity can "knock" electrons free from their shells around atomic nuclei, radioactivity is often called *ionizing radiation*. It can ionize living tissue as well as inanimate materials. Long-term exposure at low levels can cause cancer.

Short-term exposure at high levels can cause burns, sickness, and death. (Ironically, ionizing radiation in regulated doses can also be useful in the treatment of some types of cancer.) Three major types of ionizing radiation are *alpha* (α) *rays*, *beta* (β) *rays*, and *gamma* (γ) *rays*. The *nuclear decay* processes that produce these emissions are *alpha* (α) *decay*, *beta* (β) *decay*, and *gamma* (γ) *decay*.

NUCLIDES

In texts and papers dealing with nuclear decay, you'll sometimes encounter the term *nuclide*, which refers to the nucleus of a specific isotope of an element. For example:

- A C-12 nuclide is a nucleus of carbon-12, which has six protons and six neutrons.

- A C-14 nuclide is a nucleus of carbon-14, which has six protons and eight neutrons.

- A U-235 nuclide is a nucleus of uranium 235, which has 92 protons and 143 neutrons.

- A U-238 nuclide is a nucleus of uranium 238, which has 92 protons and 146 neutrons.

When a heavy nuclide decays into one or more lighter nuclides, the original nucleus is called a *parent nuclide*, and the resultant nuclei are called *daughter nuclides*.

ALPHA DECAY

When certain atomic nuclei break apart, α particles are produced. When a nucleus emits an α particle, it loses two protons and two neutrons. A parent nuclide with atomic number n and atomic weight m therefore becomes a daughter nuclide with atomic number $n - 2$ and atomic weight $m - 4$.

In nature, α decay is observed in the disintegration of U-238. A sample of this material spontaneously emits an α particle every now and then. While the timing of each individual event is random, the average frequency of events per unit time is predictable. A U-238 nuclide that emits an α particle becomes a *thorium-234* (Th-234) nuclide. The atomic number of thorium is 90, which is 2 less than the atomic number of uranium.

A barrage of α particles can be dangerous at high intensity, but the penetrating power is relatively low. A thin sheet of a heavy metal such as lead will stop most α particles.

BETA DECAY

In some atomic nuclei, a neutron can change into a proton, or vice-versa. In the former case, known as *negative beta* (β^-) *decay*, a high-speed electron (e^-) or *negative beta* (β^-) *particle* is emitted from a neutron, so the electrically neutral particle attains a unit positive charge. In the latter case, known as *positive beta* (β^+) *decay*, a high-speed positron (e^+) or *positive beta* (β^+) *particle* is emitted from a proton, so the particle with a unit positive charge becomes electrically neutral. The fact that this can occur spontaneously is interesting enough, but it's not the whole story!

If you subtract the rest mass of a proton from the rest mass of a neutron, you get a difference amounting to more than the rest mass of an electron. Conversely, if you subtract the rest mass of a neutron from that of a proton, the difference is greater (negatively) than the mass of a positron (which can be considered negative because a positron is a form of *anti-matter*). The mass defect Δm_β in a β-decay event is manifest as energy by the production of a neutrino or anti-neutrino along with the β^- or β^+ particle. The energy E_β contained in the neutrino or anti-neutrino is:

$$E_\beta = (\Delta m_\beta) \, c^2$$

This variant of the Einstein equation can serve to remind you that, whenever there is a nuclear reaction involving a mass defect, the discrepancy can be accounted for by the production or absorption of energy. If you think of matter and energy as two different phases or states of a single, more universal quantity (let's call it "mattergy"), then it's worthwhile to remember that "mattergy" can change state, but it cannot be created or destroyed.

While a single β-decay event changes the atomic number of a nucleus, the atomic weight, as it appears in a periodic table of the elements, stays the same. (The actual mass of the nucleus, expressed precisely in atomic mass units, decreases by an amount equal to the mass defect, but that is not enough to affect the atomic weight as a whole number.) In a common natural example of β^- decay, a carbon-14 (C-14) nuclide with atomic number 6 becomes a nitrogen-14 (N-14) nuclide with atomic number 7. A natural example of β^+ decay involves a neon-19 (Ne-19) nuclide with atomic number 10 turning into a fluorine-19 (F-19) nuclide with atomic number 9.

An intense barrage of β particles, like a stream of α particles, can damage living tissue because the electrons or positrons travel at speeds sufficiently great to ionize atoms they strike. Fortunately, the penetrating power of β radiation, like that of α radiation, is not great. It is easy to provide shielding against it.

GAMMA DECAY

In γ decay, a high-energy photon called a *gamma* (γ) *particle* is emitted by a radio-active nuclide. Neither the atomic number nor the atomic weight change when this happens. However, the precise mass of the nuclide after the photon is emitted, as expressed in atomic mass units, is slightly less than the mass of the nuclide before the event. The difference Δm is manifest as energy in the γ particle. Besides that, when a nuclide emits a γ particle, the nuclide *recoils* because the photon carries away some momentum. It's an atomic-scale manifestation of Newton's third law of motion.

Gamma rays, which resemble X rays but have shorter wavelength, have more penetrating power than α rays or β rays. It takes several centimeters of solid lead, or about half a meter of solid concrete, to effectively protect humans against the γ radiation from the *radioactive fallout* that could result from a nuclear explosion, reactor disaster, or "dirty bomb." For this reason, γ radiation is especially danger-ous. A burst of γ radiation from a collapsing star within a few light-years of our solar system could penetrate the earth's atmosphere and kill most living things on the planet within a few days. Fortunately, the probability of such a γ-ray burst taking place within the span of a single human lifetime is so low that we might as well not worry about it.

HALF-LIFE

The rate at which a nuclear decay process occurs is defined in terms of a time period known as the *half-life*, symbolized $t_{1/2}$. The half-life is the length of time required for half of the radioactive atoms of the parent nuclide to change into daughter nuclide(s) for a specific type of nuclear decay.

The value of $t_{1/2}$ is independent of the original quantity of radioactive material in a sample. It is also independent of the time at which observation of the sample begins. If you have a sample of radioactive material, the half-life for any specific type of decay such as α, β, or γ is a constant that can be determined by experiment. (If the sample produces more than one type of ionizing radiation, the half lives for each type differ.)

If, at a certain time, a sample of radioactive material contains n_1 radioactive atoms of the parent nuclide, then after x half-lives that same sample will contain n_2 radioactive atoms of the parent nuclide according to the following equation:

$$n_2 = n_1 / (2^x)$$

Half-lives can vary greatly depending on the parent nuclide, as well as on the form of radiation for which they are defined. For C-14, the β-decay half-life is approximately 5700 years (yr), where 1 yr = 3.15576×10^7 s. Some naturally occurring radioactive elements have half-lives of thousands or even millions of years. Certain human-made radioactive isotopes have half-lives amounting to a small fraction of 1 s. Figure 9-3 is a generic graph of radiation intensity as a function of the number of elapsed half-lives.

The quantity x in the above equation—the number of half-lives elapsed—doesn't have to be a whole number. If n_1 and n_2 are measured at two different time points t_1 and t_2, we can determine x by plugging the numbers into the above equation. We don't even have to know the actual values of n_1 and n_2; it's good enough to know the ratio n_1 / n_2, which can be inferred by taking measurements of the ionizing radiation intensities r_1 and r_2 in particles emitted per unit time (per second or per minute, for example) at t_1 and t_2. If we know these intensities, then:

$$x = \log_2 (r_1 / r_2)$$

where $\log_2$ represents the *base-2 logarithm*.

Most calculators don't find base-2 logarithms directly. But the base-2 logarithm of any number z can be found in terms of the base-10 logarithm or *common logarithm* (symbolized log). Common logarithms are dealt with by all scientific calculators. In order to find the base-2 logarithm of a number z in terms of the common logarithm, use this formula:

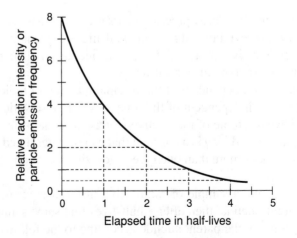

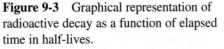

Figure 9-3 Graphical representation of radioactive decay as a function of elapsed time in half-lives.

$$\log_2 z = \log z / \log 2$$
$$= \log z / 0.301030$$
$$= 3.32193 \log z$$

When we substitute the ratio of intensities r_1 / r_2 for z in the above equation, we have, as the number of half-lives elapsed:

$$x = 3.32193 \log (r_1 / r_2)$$

If we are able to measure r_1 and r_2 with a high enough degree of precision, then the half-life of a nuclide can be found within a time frame far shorter than $t_{1/2}$ itself. That's how scientists have determined the β-decay half-life of C-14 with reasonable certainty, even though it is many human lifespans.

URANIUM FISSION

The term *fission* means "splitting apart." In *nuclear fission*, the splitting of atomic nuclei produces smaller nuclei and different elements. This can happen spontaneously over time, as we have seen. It can also be caused by the deliberate bombardment of nuclei with high-speed subatomic particles such as protons, neutrons, or α particles. In a *fission power plant*, nuclei of U-235 are split in a regulated fashion. This is called *induced fission*.

The key to producing U-235 fission is the production of a stream of high-speed neutrons. When a neutron strikes a U-235 nuclide, that nuclide splits instantly into two daughter nuclides. As this happens, two or three *secondary neutrons* are emitted along with γ rays. Thermal energy is also produced, heating up the uranium. If one of the emitted neutrons hits another U-235 nuclide, then that nuclide splits, and the process is repeated. In a large enough sample of U-235, all this fission and internal neutron emission causes a *chain reaction*. If the chain reaction goes on long enough, all the U-235 ends up getting split down into daughter nuclides. The end product is *spent nuclear fuel*.

One of three situations can prevail when U-235 is subjected to high-speed neutron bombardment. These scenarios are called the *subcritical state*, the *critical state*, and the *supercritical state*.

- In the subcritical state, the reaction dies down before much fuel is spent. This happens if, on the average, less than one emitted neutron hits a U-235 nuclide with enough energy to split it.

- In the critical state, the reaction sustains itself in a steady fashion until the fuel is spent. This happens if, on the average, one emitted neutron hits a U-235 nuclide with enough energy to split it.

- In the supercritical state, the reaction increases in intensity, and the uranium heats up uncontrollably. This happens if, on the average, more than one emitted neutron hits a U-235 nuclide with enough energy to split it.

THE FISSION POWER PLANT

In order for a nuclear fission reactor to work properly, it must be maintained in the critical state. This involves controlling the temperature of the U-235, as well as starting out with a sample having the proper mass and shape. To some extent, the amount of neutron radiation within the sample can be controlled to keep the system operating in a steady state. When regulated in just the right way, a uranium fission reaction can provide a lot of usable thermal energy for a long time.

There is no risk of explosion with U-235 that has been refined exclusively for use in nuclear reactors. If you've seen a movie in which a fission reactor blew up like an atomic bomb, it was not based on reality. Nevertheless, there is plenty to worry about if a fission reaction gets out of control. If a reactor is allowed to go supercritical for a long enough time, the uranium will melt because of the excessive heat. That's a condition called *meltdown*. The nuclear fuel can then damage the structure that contains it. In the worst case, some of the nuclear fuel can escape into the external environment, contaminating the soil, water, and air with radioactive isotopes.

A uranium fission reactor is housed in a multi-layered structure. A *radiation shield* and *containment vessel* prevent the escape of radiation or radioactive materials into the surrounding environment under normal operating conditions. The entire assembly is housed in a massive reinforced building called the *secondary containment structure*. This building, characteristically shaped like a dome or half-sphere for maximum structural integrity, is designed to withstand catastrophes such as tornadoes, hurricanes, earthquakes, and direct hits by aircraft or missiles.

THE FISSION-POWERED SHIP

Figure 9-4 is a block diagram of the power plant in a typical fission-powered ship or submarine. Heat from the reactor is transferred to a *water boiler* by means of *heat-transfer fluid*, which is similar to the fluid used in heat pumps, air conditioners, and automobile radiators. This fluid, also called *coolant*, passes from the shell of the boiler back to the reactor through a *coolant pump*. The water in the boiler is converted to steam, which drives a *turbine*. After passing through the turbine, the steam is condensed and sent back to the boiler by a *feed pump*. The water and heat-transfer fluid are entirely separate, closed systems. Neither comes into

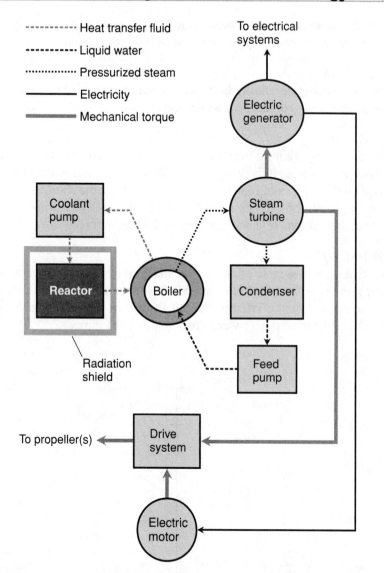

------------ Heat transfer fluid

-------- Liquid water

·············· Pressurized steam

———— Electricity

━━━━ Mechanical torque

To electrical
systems

Electric
generator

Coolant
pump

Steam
turbine

Reactor   Boiler   Condenser

Radiation
shield

Feed
pump

To propeller(s)   Drive
system

Electric
motor

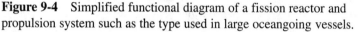

Figure 9-4   Simplified functional diagram of a fission reactor and
propulsion system such as the type used in large oceangoing vessels.

direct physical contact with the other. This prevents the accidental discharge of
radiation into the environment through the water/steam system.

The turbine can turn the vessel's propellers through a *drive system*. The turbine
also turns the shaft of a generator that provides electricity for the crew, passengers,
and electronic systems. In addition, the electricity from the generator can power a

motor, or charge a battery that powers a motor, connected to the drive system. This electric motor can provide supplemental or backup propulsion.

PROBLEM 9-6

How do γ rays compare with radio waves, infrared (IR), visible light, ultraviolet (UV) rays, and X rays in terms of wavelength?

SOLUTION 9-6

Figure 9-5 is a logarithmic graph of the electromagnetic (EM) spectrum from "longwave radio" through the γ-ray region, showing the approximate wavelength ranges for all these types of radiation. Gamma rays have the shortest wavelengths, and therefore the highest frequencies. Because of this, γ-ray photons contain more energy than the photons of any other form of EM radiation.

PROBLEM 9-7

Suppose that a given sample of radioactive material emits an average of 99 α particles per minute. After 1.0 yr has elapsed, that same sample emits an average of 90 α particles per minute. What is the half-life $t_{1/2}$ in years for α radiation from this sample? Determine the answer to two significant figures.

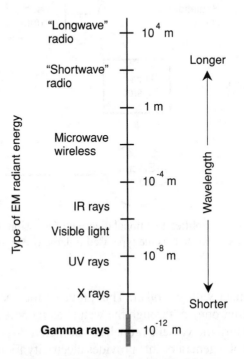

Figure 9-5 Illustration for Problem and Solution 9-6.

SOLUTION 9-7

Let r_1, the initial radiation intensity in α particles per minute, be equal to 99. Let r_2, the final radiation intensity in α particles per minute, be equal to 90. We can use the formula derived above to determine the number x of half-lives that have elapsed over the course of the year:

$$
\begin{aligned}
x &= 3.32193 \log (r_1 / r_2) \\
&= 3.32193 \log (99 / 90) \\
&= 3.32193 \log (1.1) \\
&= 3.32193 \times 0.0413927 \\
&= 0.137504
\end{aligned}
$$

That's the number of half-lives in 1.0 yr. In order to determine the number of years per half-life, we must take the reciprocal of this. Accurate to two significant figures, we obtain:

$$
\begin{aligned}
t_{1/2} &= 1.0 / 0.137504 \\
&= 7.3 \text{ yr}
\end{aligned}
$$

Nuclear Fusion

In atomic physics, the term *fusion* means "combining." In *nuclear fusion*, two or more parent nuclides merge to form a single daughter nuclide. The total mass of the daughter nuclide is less than the combined masses of the parent nuclides. The mass defect, as with fission, is manifest as energy and high-speed subatomic particles. In contrast to fission, nuclear fusion is the most common, and is easiest to obtain, with light elements such as hydrogen, helium, and carbon.

FUSION IN THE SUN

Physicists believe that the sun converts hydrogen to helium by means of fusion. *Hydrogen fusion* requires extremely high temperature. The powerful gravitation imposed by the sun's huge mass keeps the core in a constantly compressed state. This compression keeps the core hot enough for hydrogen fusion to take place continuously.

Solar hydrogen fusion is a multi-step process and is shown in Fig. 9-6. At first, two H-1 nuclides (protons) fuse, emitting a positron and a neutrino. This event is attended by a loss of a unit positive charge because one of the protons becomes a neutron. The result is the nuclide deuterium (H-2), referred to earlier in this chapter as a deuteron. It consists of one proton and one neutron. The deuteron combines

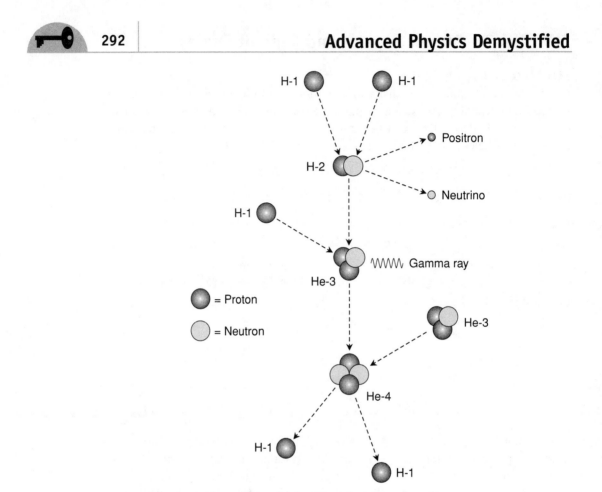

Figure 9-6   The hydrogen fusion process that takes
place in the core of the sun.

with another proton to form the nuclide *helium-3* (He-3), containing two protons
and one neutron. As this happens, a burst of γ radiation is emitted. Two He-3
nuclei, resulting from two separate iterations of the above-described process, then
combine to form the common nuclide *helium-4* (He-4), which has two protons and
two neutrons. (This is the isotope of helium we use to fill balloons.) In this final
phase, two protons are ejected. These can contribute fuel for more fusion events of
the same type.

FUSION IN BOMBS

In a *hydrogen bomb*, a different hydrogen fusion reaction takes place. This mode,
if it can ever be controlled, might be used in a *fusion reactor*. A large amount of

energy is liberated when two slightly different nuclides of *heavy hydrogen* merge. One is a deuterium, and the other is known as *tritium* (H-3), which contains one proton and two neutrons. When these combine, the result is a nucleus of He-4; the extra neutron is ejected (Fig. 9-7). Along with this, energy is liberated, just as is the case inside the sun. For this fusion mode, called *deuterium-tritium fusion* or *D-T fusion*, ordinary hydrogen (H-1) nuclides, consisting of individual protons, won't work directly. Several other fuel combinations can theoretically be used to obtain nuclear fusion, but the D-T mode has received the most attention.

In the sun, the fusion process goes on continuously because of the heat produced by the crushing pressure of gravitation, and also because of the heat generated from the reactions themselves. In a hydrogen bomb, heat to start the reaction is supplied by a fission bomb, but the reaction burns itself out in a hurry. It's impossible to get enough gravitation to start and maintain fusion indefinitely and in a controlled manner using a human-made system in an earthbound environment. In order to make fusion work for the purpose of generating useful power, the fuel must somehow be confined.

PLASMA FUEL

When a sample of gas is heated to an extremely high temperature, the electrons are stripped away from the nuclei, forming positive ions and free electrons. When this ionization occurs, the gas, which is normally a poor conductor of electric current,

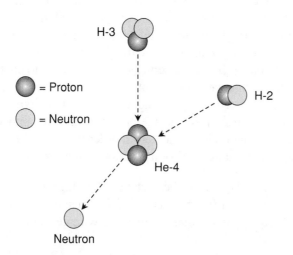

Figure 9-7 The fusion process that occurs in a hydrogen bomb.

becomes a good conductor. A substance in this state is known as a *plasma*. Because a plasma differs from an ordinary gas, the plasma state has been called the *fourth phase of matter* (the other three being solid, liquid, and gas). In the prototypes of fusion reactors that most scientists favor, deuterium and tritium exist in the plasma state, heated to temperatures comparable to those in the center of the sun.

External electric (E) or magnetic (B) fields can cause a plasma to constrict, distort, bunch up, or spread out. If a plasma is surrounded by an external E field or B field having certain properties, the plasma can be kept within a small, defined space, even if it becomes hot enough to sustain hydrogen fusion reactions. The external fields act on the plasma in much the same way as gravitation inside the sun keeps the hot core gases confined. The use of external B fields to compress and hold a hot D-T plasma in place during a fusion reaction is known as *magnetic confinement*.

THE TOKAMAK

The most promising method of magnetic confinement makes use of an evacuated toroidal (donut-shaped) enclosure called a *tokamak*. This term is an acronym derived from a Russian descriptive phrase that translates as "toroidal chamber and magnetic coil." The plasma is contained inside the chamber.

In the tokamak (Fig. 9-8), two sets of coils, called the *toroidal field coils* and the *poloidal field coils*, surround the enclosure. The coils carry electric currents that produce strong B fields. An electric current of up to 5×10^6 A, provided by a large transformer, travels through the plasma around the toroid in a circular, endless loop. This plasma current creates a B field of its own. The B fields from the currents in the coils and the plasma interact, confining the plasma, aligning it within the chamber and forcing it toward the center of the chamber cross-section, keeping it away from the walls. This is important, because the plasma must be heated to more than 10^8 K for fusion to occur. If such superheated plasma were to contact the tokamak wall at any point, the chamber would rupture, air would leak in, the plasma would cool below the critical temperature, and the fusion reaction would cease. The helium product of the fusion process is removed from the chamber by *divertors*.

Several processes can be implemented in order to obtain the high plasma temperature necessary to sustain the fusion reaction in the tokamak:

- *Ohmic heating* arises from the fact that the current, as it circulates in the plasma, encounters a finite resistance. This means that a certain amount of power is dissipated in the plasma, just as a wire gets hot when it carries high current.

- *Self heating* takes place because the fusion reaction produces thermal energy, and some of this heat is absorbed by the plasma.

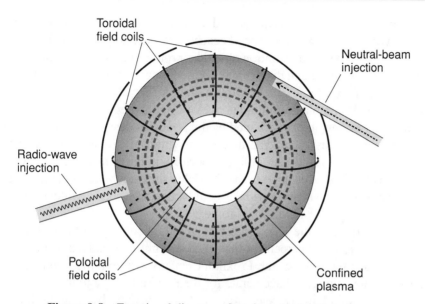

Figure 9-8 Functional diagram of a tokamak, showing plasma confinement coils and two methods of heating the plasma.

- *Neutral-beam injection* involves firing high-energy beams of neutral H-2 and H-3 atoms into the plasma. These heat the plasma when they collide with its atoms. The injected atoms must be electrically neutral so they can penetrate the B fields inside the tokamak without being deflected and striking the chamber walls.

- *Radio-wave injection* is done by transmitting EM waves into the plasma at several points in the chamber. These waves have a frequency such that their energy is absorbed by the plasma. Figure 9-8 shows one point of neutral-beam injection and one point of radio-wave injection.

FUSION FOR SPACECRAFT PROPULSION

Several types of fusion-powered spacecraft have been proposed by scientists and aerospace engineers as alternatives for obtaining the speeds necessary for long-distance space journeys, using fuel that would be of reasonable mass. These include designs known as the *Orion*, the *Daedalus*, and the *Bussard Ramjet*.

In the Orion space ship, hydrogen fusion bombs would be exploded at regular intervals to drive the vessel forward. The force of each blast, properly deflected, would accelerate the ship. The blast deflector would be strong enough to withstand the violence of the bomb explosions, and it would be made of material that would not melt, vaporize, deform, or erode because of the explosions. The blast deflector would also serve as a radiation shield to protect the astronauts in the living quarters. This idea has been discounted in recent years because of technical problems, such as the extreme bursts of acceleration (harmful to the crew) and the danger that the bomb blasts could damage or disable the ship.

A smoother ride would be provided by the Daedalus design. This would replace the bombs with a fusion reactor that would produce a continuous, controlled "burn." This vessel would use a blast deflector similar to that used in the Orion design. The principal advantage of the Daedalus ship would lie in the fact that the acceleration would be steady. Daedalus could attain high speed without subjecting the astronauts to bursts of extreme g-force.

Either the Orion or the Daedalus ships could, according to their proponents, reach approximately 10 percent of the speed of light ($0.1c$). This would make it possible to reach the nearest star, Proxima Centauri, within a lifespan of a human on the ship. The ship would accelerate for the first half of the journey, and decelerate for the second half. Deceleration would be obtained by rotating the vessel 180° so the exhaust would be directed forward, producing a rearward impulse.

The most intriguing nuclear fusion design is the Bussard Ramjet. This is similar to the Daedalus, but it would not have to carry nearly as much fuel. Once the ship got up to a certain speed, a huge "scoop" in the front could (hopefully!) gather up enough hydrogen atoms from interstellar space to provide the necessary fuel for hydrogen fusion reactions. The greater the speed attained, the more hydrogen the ship could sweep up, thus helping it go even faster. The ship would decelerate by directing the exhaust out of a forward-facing nozzle in the center of the scoop.

PROBLEM 9-8

Will the powerful magnetic fields produced by the field coils and plasma in a tokamak pose a danger to the people who operate and maintain it?

SOLUTION 9-8

This should not be a problem. The reactor and its housing can be surrounded with a ferromagnetic metal enclosure to keep the magnetic fields away from personnel in the vicinity.

PROBLEM 9-9

Could a hydrogen-fusion space vessel be launched directly from the earth's surface? If so, wouldn't the blast from the engines cause destruction and deadly radiation near the launch site?

SOLUTION 9-9

Most proposals for hydrogen-fusion-powered spacecraft envision a conventional rocket launch vehicle to place the ship in an earth orbit. The fusion engines would be activated at an altitude of several thousand kilometers. There would be no dangerous effects on the earth's surface.

Quiz

This is an "open book" quiz. You may refer to the text in this chapter. A good score is 8 correct. Answers are in the back of the book.

1. Suppose a parent nuclide with atomic number n and atomic weight m emits a γ particle. What are the atomic number and atomic weight of the daughter nuclide?

    (a) Atomic number n, atomic weight m.

    (b) Atomic number $n + 1$, atomic weight $m - 1$.

    (c) Atomic number $n - 1$, atomic weight m.

    (d) Atomic number n, atomic weight $m - 1$.

2. An atom of carbon with five protons rather than the usual six is an example of

    (a) a positive carbon ion.

    (b) a negative carbon ion.

    (c) an isotope of carbon.

    (d) a mistake in somebody's thinking, because carbon, whose atomic number is 6, always has atoms with six protons in the nucleus.

3. Consider the β-decay half-life of a certain material to be 6.755×10^5 s. Imagine that you have a sample of that material in your lab right now. After how long will the rate of β-decay of this material be reduced to 10 percent of its present rate?

    (a) 2.244×10^6 s.

    (b) 3.322×10^6 s.

    (c) 6.755×10^6 s.

    (d) More information is necessary to answer this question.

4. In the Balmer formula for hydrogen gas, as the principal quantum number representing the shell from which an electron falls becomes larger and larger without limit, the wavelength decreases, converging on a lower limit of

   (a) about 1100 nm.

   (b) about 365 nm.

   (c) about 91 nm.

   (d) zero.

5. Suppose a parent nuclide with atomic number n and atomic weight m emits a high-speed positron. What are the atomic number and atomic weight of the daughter nuclide?

   (a) Atomic number n, atomic weight m.

   (b) Atomic number $n + 1$, atomic weight $m - 1$.

   (c) Atomic number $n - 1$, atomic weight m.

   (d) Atomic number n, atomic weight $m - 1$.

6. In the tokamak, the nuclear fuel is confined by means of

   (a) gravitation.

   (b) radio waves.

   (c) magnetic fields.

   (d) high-speed neutrons.

7. Suppose that a small ball massing 32 g is struck with a club and propelled at an initial linear speed of 75 m/s. The deBroglie wavelength of this moving object, in theory, is

   (a) longer than the wavelength of visible light.

   (b) in the visible-light range.

   (c) shorter than the wavelength of visible light.

   (d) meaningless, because in order to calculate it we are forced to divide by zero.

8. Which of the following types of ionizing radiation has the most penetrating power?

   (a) α rays.

   (b) β rays.

   (c) γ rays.

   (d) All three of the above types of radiation have the same penetrating power.

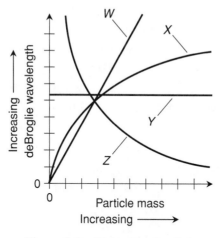

Figure 9-9   Illustration for Quiz Question 10.

9. An electron can gain energy within an atom, thereby moving to a larger shell,

   (a) when that atom absorbs a photon having a certain frequency.

   (b) when that atom emits a photon having a certain frequency.

   (c) when that electron becomes more negatively charged.

   (d) when that electron becomes less negatively charged.

10. Consider a stream of particles of various rest masses, all of which are moving at the same speed. Suppose you plot a graph of the deBroglie wavelengths as a function of particle rest mass. Which of the plots in Fig. 9-9 best represents this function? Assume that both graph scales are linear.

   (a) Line *W*.

   (b) Curve *X*.

   (c) Line *Y*.

   (d) Curve *Z*.

Final Exam

Do not refer to the text when taking this exam. A good score is at least 75 correct. Answers are in the back of the book. It's best to have a friend check your score the first time, so you won't memorize the answers if you want to take the exam again.

1. Imagine a satellite in a circular orbit around the earth, passing precisely over the geographic poles. Suppose the radius of the orbit is 1.000×10^4 km. Imagine a plane P that contains the equator and extends indefinitely into space. When the satellite is over a point Q on the earth's surface at 60° north latitude, how far from plane P is it, as measured along a straight line perpendicular to P?

   (a) 8660 km.

   (b) 7071 km.

   (c) 6667 km.

   (d) 5000 km.

   (e) More information is needed to answer this.

2. Assuming the acceleration of gravity exerted on the satellite in the scenario of Question 1 is 9.8 m/s^2, how long does it take the satellite to pass from plane P to a point directly over point Q as it travels northward?

   (a) 3000 s.
   (b) 3600 s.
   (c) 3345 s.
   (d) 4500 s.
   (e) More information is needed to answer this.

3. Let C represent the temperature in degrees Celsius and F represent the temperature in degrees Fahrenheit. Which of the following formulas describes how a Fahrenheit reading can be converted to a Celsius reading?

   (a) $C = (9/5)F + 32$.
   (b) $C = (5/9)F - 32$.
   (c) $C = (5/9)(F - 32)$.
   (d) $C = (9/5)(F + 32)$.
   (e) None of the above

4. Which of the following (a), (b), (c), or (d), if any, is *not* an example of an accelerating object?

   (a) A car speeding up while driving on a straight, level highway.
   (b) A space ship slowing down while traveling in a straight line.
   (c) A car moving at constant speed around a curve.
   (d) An artificial satellite orbiting the moon.
   (e) All of the above (a), (b), (c), and (d) are examples of accelerating objects.

5. Consider an object in a circular orbit around the earth at an altitude such that the earth's gravitational force vector on a mass of 10 kg has a magnitude of 90 N. What is the acceleration vector of this object?

   (a) 3.0 m/s^2 away from the earth's center.
   (b) 3.0 m/s^2 in the same direction as the instantaneous velocity vector.
   (c) 3.0 m/s^2 toward the earth's center.
   (d) 3.0 m/s^2 in the direction opposite the instantaneous velocity vector.
   (e) None of the above

6. The total quantity of electric flux that passes through any closed, non-conducting surface surrounding a point charge is always equal to the charge quantity

(a) times the permittivity of free space.

(b) divided by the permittivity of free space.

(c) times the permittivity of the medium inside the enclosure.

(d) divided by the permittivity of the medium inside the enclosure.

(e) divided by the surface area of the enclosure.

7. Figure Exam-1 shows a collision between two objects initially moving in straight lines, and neither of which is rotating before they hit. The masses and velocities are as follows:

$$m_1 = \text{mass of object 1}$$
$$m_2 = \text{mass of object 2}$$
$$\mathbf{v}_{11} = \text{velocity of object 1 before collision}$$
$$\mathbf{v}_{21} = \text{velocity of object 2 before collision}$$
$$\mathbf{v}_{12} = \text{velocity of object 1 after collision}$$
$$\mathbf{v}_{22} = \text{velocity of object 2 after collision}$$

Which of the following equations is true in general?

(a) $(m_1\mathbf{v}_{11})^2 + (m_2\mathbf{v}_{21})^2 = (m_1\mathbf{v}_{12} + m_2\mathbf{v}_{22})^2$.

(b) $m_1\mathbf{v}_{11} + m_2\mathbf{v}_{21} = m_1\mathbf{v}_{12} + m_2\mathbf{v}_{22}$.

(c) $m_1\mathbf{v}_{11} - m_2\mathbf{v}_{21} = m_1\mathbf{v}_{12} - m_2\mathbf{v}_{22}$.

(d) $(m_1\mathbf{v}_{11})^2 - (m_2\mathbf{v}_{21})^2 = (m_1\mathbf{v}_{12} - m_2\mathbf{v}_{22})^2$.

(e) None of the above

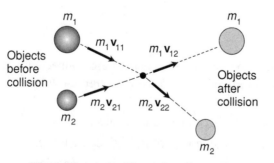

Figure Exam-1 Illustration for Exam Question 7.

8. If the inductance of a component increases by a factor of 8, then its inductive reactance at a constant frequency

   (a) decreases by a factor of 64.

   (b) decreases by a factor of 8.

   (c) stays the same.

   (d) increases by a factor of 8.

   (e) increases by a factor of 64.

9. For an object moving in a straight line with acceleration-vs.-time function $\mathbf{a}(t)$, velocity-vs.-time function $\mathbf{v}(t)$, and displacement-vs.-time function $\mathbf{q}(t)$, we can derive $|\mathbf{q}|(t)$ on the basis of $|\mathbf{a}|(t)$ as follows:

$$|\mathbf{q}|(t) = \int \int |\mathbf{a}|(t)\, dt\, dt$$

If the initial time is assigned the value $t = 0$, then the constants of integration in this formula are determined by the

   (a) initial value of $|\mathbf{a}|$.

   (b) initial value of $|\mathbf{v}|$.

   (c) initial value of $|\mathbf{q}|$.

   (d) initial values of $|\mathbf{v}|$ and $|\mathbf{q}|$.

   (e) initial values of $|\mathbf{a}|$ and $|\mathbf{v}|$.

10. Consider two complex impedances represented as vectors $\mathbf{Z}_1$ and $\mathbf{Z}_2$ at an AC frequency f, as follows:

$$\mathbf{Z}_1 = 5 + j7$$
$$\mathbf{Z}_2 = 2 - j7$$

If these two complex impedances are connected in series and an AC current having frequency f passes through the whole circuit, then that circuit will act as

   (a) a pure, finite, non-zero resistance.

   (b) a pure, finite, non-zero inductive reactance.

   (c) a pure, finite, non-zero capacitive reactance.

   (d) a theoretically perfect short circuit.

   (e) a theoretically perfect open circuit.

11. Consider two complex admittance vectors:

$$\mathbf{Y}_1 = G_1 + jB_1$$
$$\mathbf{Y}_2 = G_2 + jB_2$$

What is the net admittance vector $\mathbf{Y}$ when components having these admittances are connected in parallel?

(a) $\mathbf{Y} = j[(G_1 + G_2)(B_1 + B_2)]^{1/2}$.

(b) $\mathbf{Y} = j(G_1 + G_2)^2 (B_1 + B_2)^2$.

(c) $\mathbf{Y} = j(G_1 + G_2) (B_1 + B_2)$.

(d) $\mathbf{Y} = (G_1 + G_2) + j(B_1 + B_2)$.

(e) None of the above

12. Mechanical work can be expressed as

(a) the integral of potential energy over time.

(b) the integral of kinetic energy over time.

(c) the derivative of potential energy with respect to time.

(d) the derivative of kinetic energy with respect to time.

(e) None of the above

13. Imagine two objects in free fall in the earth's gravitational field. One object has a mass of 10 kg and the other has a mass of 20 kg. Therefore, the gravitational force on the less massive object is

(a) one-quarter of the gravitational force on the more massive object.

(b) half the gravitational force on the more massive object.

(c) the same as the gravitational force on the more massive object.

(d) twice the gravitational force on the more massive object.

(e) four times the gravitational force on the more massive object.

14. The root-mean-square (rms) voltage of an AC wave is

(a) the square root of the peak-to-peak voltage times the square of the average voltage.

(b) the square root of the peak voltage times the square of the average voltage.

(c) the square root of the average of the squares of all the instantaneous voltages.

(d) the square of the peak voltage times the square root of the average voltage.

(e) the square of the peak-to-peak voltage times the square root of the average voltage.

15. The fundamental unit of work, the newton-meter, is theoretically equivalent to

    (a) the fundamental unit of power, the watt.

    (b) the fundamental unit of power, the watt per second.

    (c) the fundamental unit of energy, the joule.

    (d) the fundamental unit of energy, the joule per second.

    (e) None of the above

16. Fill in the blank in the following sentence to make it true: "The _____ of the atoms or molecules in a dry gas is directly proportional to the absolute thermodynamic temperature."

    (a) mean diameter

    (b) mean rotational period

    (c) mean angular frequency

    (d) mean emission wavelength

    (e) mean kinetic energy

17. Imagine a point mass P having a mass of 4.00 g revolving in a perfect circle at constant tangential speed of 25.0 m/s around a central point, as shown in Fig. Exam-2. What is the rotational kinetic energy of P?

    (a) 0.0500 J.

    (b) 0.100 J.

    (c) 1.25 J.

    (d) 2.50 J.

    (e) More information is needed to answer this.

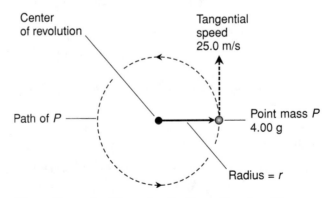

Figure Exam-2   Illustration for Exam Question 17.

18. The product of the complex numbers $3 + j6$ and $3 - j6$ is
   (a) $9 - j36$.
   (b) $9 + j36$.
   (c) $j36$.
   (d) $-j36$.
   (e) 45.

19. For a rotating solid disk with rotational inertia I and angular speed ω, the angular momentum magnitude I is the product of the rotational inertia and the angular speed:

$$L = I\omega$$

where L is in kilogram meters squared per second, I is in kilogram meters squared, and ω is in
   (a) radians per second.
   (b) revolutions per second.
   (c) degrees per second.
   (d) meters per second.
   (e) kilogram meters per second.

20. The period T_x for an object X in a circular orbit of radius s around a central object Y of mass m_y is:

$$T_x = 2\pi \, [s^3 / (Gm_y)]^{1/2}$$

where T_x is in seconds, s is in meters, G is the gravitational constant which has a value of about 6.67×10^{-11}, and m_y is in kilograms. If the orbit of object X is an ellipse that is not a perfect circle, then the radius s in the above equation must be replaced by
   (a) the length of the major axis of the ellipse.
   (b) the length of the minor axis of the ellipse.
   (c) half the length of the major axis of the ellipse.
   (d) half the length of the minor axis of the ellipse.
   (e) the difference between the lengths of the major and minor axes of the ellipse.

21. Suppose a resistor, inductor, and capacitor are connected in parallel. Suppose the resistor has a conductance of 0.050 S, the inductive susceptance is $-j0.070$ S, and the capacitive susceptance is $j0.070$ S at a frequency of 21 MHz. What can we say about this circuit without making any calculations?

   (a) It acts as a short circuit at 21 MHz.

   (b) It acts as an open circuit at 21 MHz.

   (c) Its net reactance is $j0.050$ Ω at 21 MHz.

   (d) It is parallel-resonant at 21 MHz.

   (e) None of the above

22. Imagine that you're in a classroom and the professor says:

   "Here is an expression for the increase in potential energy E_p acquired by an object of mass m that travels straight upward from altitude h_1 to altitude h_2:

   $$E_p = 9.807m\,(h_2 - h_1)$$

   where E_p is in joules, 9.807 is the acceleration of gravity in meters per second squared, m is in kilograms, and h_1 and h_2 are in meters."

   In order to make this statement more complete, the professor would do well to add,

   (a) "This formula applies only if the object does not rise so far that the gravitational acceleration on it diminishes during its journey."

   (b) "This formula applies only for an object having a large enough mass to produce a significant gravitational field of its own."

   (c) "This formula applies only for an object in a closed orbit around the earth."

   (d) "This formula applies only if we measure the altitudes h_1 and h_2 with respect to the earth's center of mass."

   (e) All of the above

23. Suppose a point charge of quantity q exists at the center of a non-conducting sphere. Also suppose that the permittivity is uniform throughout the interior of the sphere, as well as in its vicinity. Under these conditions, what might cause the E-field magnitude to vary from place to place on the surface of the sphere?

   (a) A change in the permittivity of the medium inside the sphere.

   (b) A change in the radius of the sphere.

   (c) A change in the temperature inside the sphere.

   (d) A change in the pressure inside the sphere.

   (e) Movement of the point charge away from the center of the sphere.

24. The minimum escape speed for an object from the surface of a planet having a radius of 6000 km depends on

(a) the mass of the object.

(b) the radius of the object.

(c) the mass of the planet.

(d) the distance of the planet from its sun.

(e) All of the above

25. Imagine that you're in an engineering class, and the professor gives you the following instructions for finding the complex impedance vector **Z** of a circuit containing a resistor, an inductor, and a capacitor in parallel at a certain frequency. You are, of course, told the values of the components, and also the frequency.

- Find the real-number conductance $G = 1/R$ of the resistor.
- Find the imaginary-number susceptance jB_L of the inductor.
- Find the imaginary-number susceptance jB_C of the capacitor.
- Find the net imaginary-number susceptance $jB = jB_L + jB_C$.
- Determine the real-number value $G^2 + B^2$.
- Compute the real-number values R and X in terms of the real-number values G and B, paying careful attention to the signs, using these formulas:

$$R = G / (G^2 + B^2)$$
$$X = -B / (G^2 + B^2)$$

- Finally, put together the complex impedance vector $\mathbf{Z} = R + jX$ from the values of R and X you just found.

What, if anything, is wrong with these instructions?

(a) Nothing.

(b) Conductance is an imaginary-number quantity, not a real-number quantity.

(c) Susceptance is a real-number quantity, not an imaginary-number quantity.

(d) The equations are meaningless because the quantity $G^2 + B^2$ is always equal to zero in a parallel circuit.

(e) It is impossible to find the complex impedance vector of a circuit containing a resistor, an inductor, and a capacitor in parallel at any frequency.

26. Expressed in base SI units to three significant figures, exactly one British thermal unit of heat is the equivalent of

    (a) 1.06×10^3 kg · m^2/s^2.

    (b) 1.06×10^{-3} kg · m/s^2.

    (c) 0.239 kg · m^2/s^2.

    (d) 9.48×10^{-4} kg · m/s.

    (e) 4.18 kg · m/s^2.

27. Fill in the blank to make the following sentence true: "For a particle revolving around a central point in uniform circular motion, the magnitude of the angular acceleration vector is zero, and the _____ is non-zero and constant."

    (a) magnitude of the centripetal acceleration vector

    (b) direction of the tangential velocity vector

    (c) direction of the centripetal acceleration vector

    (d) magnitude of the angular displacement vector

    (e) direction of the linear displacement vector

28. Imagine that you are on a ship coasting through interplanetary space on the way to Mars. You discover that an immediate course correction is necessary. The computers indicate that the vessel must gain 40 m/s of forward speed to get back on course. Suppose the mass of the craft is 10,000 kg. The rockets produce a constant force of 20,000 N. What is the impulse that produces the required course correction?

    (a) 4.0×10^5 kg · m/s.

    (b) 2.0×10^5 kg · m/s.

    (c) 8.0×10^4 kg · m/s.

    (d) 4.0×10^4 kg · m/s.

    (e) More information is necessary to answer this.

29. Two factors determine the angular frequency at which an ideal spring-loaded system of the sort shown in Fig. Exam-3 will oscillate: the mass m of the object attached to the movable end of the spring and the

    (a) peak displacement q_{pk+} or q_{pk-} from the equilibrium point.

    (b) spring constant.

    (c) acceleration of gravity.

    (d) overall length of the spring at the equilibrium point.

    (e) initial force applied to the mass to cause the spring to compress or stretch.

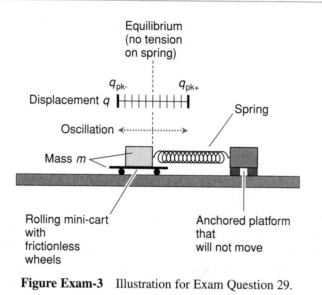

Figure Exam-3 Illustration for Exam Question 29.

30. Imagine that you stand on the top of a high mountain on a planet with no atmosphere (so there is no air resistance) where the acceleration of gravity is 1.10 m/s^2. Suppose you aim a rifle in a precisely horizontal direction and then fire a bullet from it. What is the magnitude of the vertical component of the bullet velocity vector after 2.00 s, assuming the bullet does not hit the surface or any other obstruction?

(a) It depends on the mass of the bullet.

(b) It depends on the initial horizontal speed of the bullet.

(c) 1.10 m/s.

(d) 1.21 m/s.

(e) 2.20 m/s.

31. In the scenario of Question 30, let **v** represent the instantaneous velocity vector of the bullet, and let **a** represent its instantaneous acceleration vector. An infinitesimally short time after the bullet is fired,

(a) dir **a** is perpendicular to dir **v**.

(b) dir **a** is at a $45°$ angle with respect to dir **v**.

(c) dir **a** is the same as dir **v**.

(d) dir **a** is horizontal and dir **v** is vertical.

(e) None of the above

32. In the scenario of Question 30, suppose the mass of the bullet is 10.5 g and the initial horizontal speed of the bullet is 1132 m/s. What is the kinetic energy contained by the bullet an infinitesimally short time after it is fired?

    (a) 5.94 J.

    (b) 11.9 J.

    (c) 6.73 kJ.

    (d) 13.5 kJ.

    (e) It depends on the horizontal acceleration of the bullet.

33. Imagine an extremely thin, flat, square sheet of metal that measures precisely 1 m along each edge when the temperature is 270 K. Suppose the coefficient of linear expansion for this material is $+2.00 \times 10^{-6}$ /K. If the absolute temperature rises from 270 K to 280 K, by what percentage does the surface area of this sheet increase? Express the answer to three significant figures.

    (a) 4.00%.

    (b) 0.400%.

    (c) 0.0400%.

    (d) 0.00400%.

    (e) More information is needed to answer this.

34. Suppose a single electronic component exhibits a pure imaginary-number inductive reactance of $j5.00$ Ω at a frequency of 16.5 MHz. This is equivalent to a pure imaginary-number inductive susceptance at 16.5 MHz of

    (a) $j200$ mS.

    (b) $-j200$ mS.

    (c) $200 + j200$ mS.

    (d) $200 - j200$ mS.

    (e) $200 + j0.00$ mS.

35. The Schwarzchild radius of a collapsing star is the radius at which

    (a) all of the electrons are driven into the atomic nuclei.

    (b) all the atoms become crushed together, forming neutrons.

    (c) nuclear fusion can no longer take place, and the star goes out.

    (d) the gravitational field strength becomes infinite.

    (e) the minimum escape speed reaches the speed of light in free space.

36. Suppose two objects at different temperatures are placed in direct contact, and thermal energy is allowed to flow freely between them. Eventually, if no thermal energy is supplied from the outside, the temperatures of the two objects will become equal. What is the technical term for this process?

    (a) Thermal equilibrium.

    (b) Heat transfer equalization.

    (c) Proportional heat transfer.

    (d) Heat media transfer.

    (e) None of the above

37. Fill in the blank to make the following sentence true: "For an AC sine wave, the rms amplitude is equal to 0.707 times the peak amplitude, and _____ the peak-to-peak amplitude."

    (a) twice

    (b) 1.414 times

    (c) half

    (d) 0.354 times

    (e) the same as

38. Fill in the blank to make the following sentence true: "In an electrical transmission line, Z_o represents the _____, assuming the line is infinitely long and is fed with AC at one end."

    (a) power dissipated in watts per meter

    (b) strength of the magnetic field in webers

    (c) ratio of the voltage to the current in volts per ampere

    (d) strength of the electric field in volts per meter

    (e) resistance in ohms per meter

39. The density of an object at its Schwarzchild radius

    (a) varies in direct proportion to the square of the mass.

    (b) varies in direct proportion to the mass.

    (c) varies in direct proportion to the square root of the mass.

    (d) varies in inverse proportion to the mass.

    (e) varies in inverse proportion to the square of the mass.

40. In a collision between two objects both moving in straight lines before they hit, linear momentum is conserved as long as neither object is rotating initially and

   (a) both of the objects absorb all the impact when the collision takes place.

   (b) the total mass remains constant and no outside force is imposed.

   (c) the paths of the objects are perpendicular to each other prior to the collision.

   (d) neither of the objects is at rest prior to the collision.

   (e) neither of the objects is at rest after the collision.

41. In Fig. Exam-4, which of the plots portrays an isobaric process?

   (a) Line W.

   (b) Line X.

   (c) Line Y.

   (d) Curve Z.

   (e) None of them.

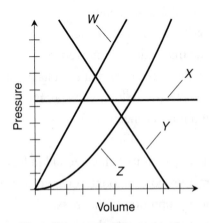

Figure Exam-4 Illustration for Exam Question 41.

42. When a point charge X is near two or more other point charges, the net electric force vector on X that is imposed collectively by all the other point charges

    (a) is impossible to predict in theory; it can only be determined by experimentation.

    (b) is equal to the vector average of the force vectors produced by each of the other point charges individually.

    (c) is equal to the vector sum of the force vectors produced by each of the other point charges individually.

    (d) is equal to the dot product of the force vectors produced by each of the other point charges individually.

    (e) is equal to the cross product of the force vectors produced by each of the other point charges individually.

43. Imagine a solid, massive disk on a frictionless bearing. Suppose that a constant force vector **F** of magnitude 10 N is applied at a point P that is 20 cm from the center of the disk. Also suppose that dir **F** lies in the same plane as the disk and is perpendicular to a radius vector **r** extending from the center of the disk to point P. This causes the disk to rotate in its own plane like a wheel, faster and faster. You observe the disk "face-on" from a point on its axis of rotation but far away from it, and from this perspective, you see the disk rotate clockwise. Let τ be the torque vector produced by **F**. Then dir τ is

    (a) exactly the same as dir **F**, rotating clockwise at the same rate as **F**.

    (b) directly away from you, at a right angle to the plane of the disk.

    (c) directly towards you, at a right angle to the plane of the disk.

    (d) perpendicular to dir **F**, but in the plane of the disk and rotating clockwise at the same rate as **F**.

    (e) exactly opposite dir **F**, rotating clockwise at the same rate as **F**.

44. In the scenario of Question 43, what is $|\tau|$?

    (a) 0.20 N · m.

    (b) 0.40 N · m.

    (c) 0.50 N · m.

    (d) 2.0 N · m.

    (e) 50 N · m.

45. Suppose you throw a golf ball horizontally outward from the base of a cliff that overlooks a flat valley whose floor is 2000 m below the edge of the cliff. Neglecting the effects of air resistance and your own height, how long after you release the ball will it strike the valley floor? Consider the acceleration of gravity to be 9.80 m/s^2.

    (a) It depends on the initial horizontal velocity component.

    (b) 408 s.

    (c) 204 s.

    (d) 20.2 s.

    (e) 10.1 s.

46. In the scenario of Question 45, how fast will the ball be traveling when it hits the valley floor?

    (a) It depends on the initial horizontal velocity component.

    (b) 99 m/s.

    (c) 198 m/s.

    (d) 4.12 km/s.

    (e) 8.24 km/s.

47. Suppose a charged particle travels through a magnetic field. The instantaneous magnetic force vector magnitude on that particle is zero

    (a) when its instantaneous velocity vector is parallel to the magnetic lines of flux.

    (b) when its instantaneous velocity vector is constant.

    (c) when its instantaneous acceleration vector is parallel to the magnetic lines of flux.

    (d) when its instantaneous velocity vector is perpendicular to the magnetic lines of flux.

    (e) under no circumstances.

48. Which, if any, of the following could theoretically be used as a unit of weight?

    (a) The kilogram-meter per second.

    (b) The foot-pound.

    (c) The gram-centimeter per second squared.

    (d) The joule-meter per second.

    (e) None of the above

49. The principal quantum number is an indicator of
    (a) the number of electrons in an atom.
    (b) the number of protons in an atom.
    (c) the number of neutrons in an atom.
    (d) the total number of particles in an atom.
    (e) None of the above

50. Suppose that a total electric flux of 5.000 N · m²/C passes at a 45.00° angle through a flat sheet of non-conducting material. If the flux lines are straight and parallel, and the sheet is turned so the flux lines pass through it at a perfect right angle, how much total flux will pass through the sheet then?
    (a) 2.500 N · m²/C.
    (b) 3.536 N · m²/C.
    (c) 5.000 N · m²/C.
    (d) 7.071 N · m²/C.
    (e) 10.00 N · m²/C.

51. The magnitude of the gravitational force between two objects is directly proportional to the product of their masses, and is inversely proportional to the
    (a) square root of the distance between their centers of mass.
    (b) distance between their centers of mass.
    (c) 3/2 power of the distance between their centers of mass.
    (d) square of the distance between their centers of mass.
    (e) cube of the distance between their centers of mass.

52. Figure Exam-5 shows two electrically charged objects X and Y close to each other, along with the electric flux lines and the direction of the flux flow in their vicinity. Based on the information in this drawing, what can be said with certainty about the electrical polarities of the objects?
    (a) The polarity of X is positive and the polarity of Y is negative.
    (b) The polarity of X is negative and the polarity of Y is positive.
    (c) The polarities of X and Y are both positive.
    (d) The polarities of X and Y are both negative.
    (e) There isn't enough information here to draw a definite conclusion about the electrical polarities of the objects.

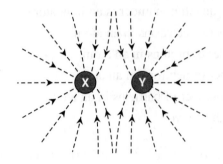

Figure Exam-5 Illustration for Exam Question 52.

53. In a series-resonant *RLC* circuit,

   (a) the net reactance is theoretically infinite.

   (b) the net reactance is theoretically zero.

   (c) the net reactance is finite, non-zero, and purely inductive.

   (d) the net reactance is finite, non-zero, and purely capacitive.

   (e) None of the above

54. An object that contains one proton, one neutron and no electrons could be called

   (a) an alpha particle.

   (b) a deuteron.

   (c) a triton.

   (d) a lithium isotope.

   (e) a lithium ion.

55. What is the total charge quantity carried by all the electrons in the shells of an electrically neutral atom of Neon-19 (Ne-19), which has an atomic number of 10?

   (a) −10 ecu.

   (b) +10 ecu.

   (c) −19 ecu.

   (d) +19 ecu.

   (e) More information is needed to answer this.

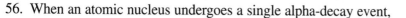

56. When an atomic nucleus undergoes a single alpha-decay event,
    (a) its atomic number and atomic weight both decrease by 2.
    (b) its atomic number stays the same, but its atomic weight decreases by 2.
    (c) its atomic number decreases by 2, but its atomic weight stays the same.
    (d) its atomic number and atomic weight both stay the same.
    (e) its atomic number decreases by 2, and its atomic weight decreases by 4.

57. Suppose you lift a 3.50-kg block for a distance of 2.30 m in a straight, vertical line using a rope and pulley system in which the pulley has no friction and the rope has no mass. How much work is expended if the apparatus is located on a planet where the acceleration of gravity is 12.0 m/s²?
    (a) 8.05 N · m.
    (b) 27.6 N · m.
    (c) 42.0 N · m.
    (d) 96.6 N · m.
    (e) 222 N · m.

58. Suppose two atomic nuclei fuse and the mass of the new nuclei is less than the sum of the masses of the two original nuclei by 1.111×10^{-25} g. How much energy (to three significant figures) is liberated or absorbed in this process?
    (a) 3.33×10^{-20} J is liberated.
    (b) 3.33×10^{-20} J is absorbed.
    (c) 1.00×10^{-11} J is liberated.
    (d) 1.00×10^{-11} J is absorbed.
    (e) More information is needed to answer this.

59. Newton's second law for rotational motion is the circular-motion counterpart of Newton's second law for linear motion, which states that the magnitude of the linear force vector on an object moving in a straight line is equal to the product of the mass of the object and the magnitude of its linear acceleration vector. How can Newton's second law for rotational motion be stated in similar terms?
    (a) The magnitude of the torque vector on a solid rotating object is equal to its mass divided by the magnitude of its angular acceleration vector.
    (b) The magnitude of the torque vector on a solid rotating object is equal to the product of its mass and the magnitude of its angular acceleration vector.

(c) The magnitude of the torque vector on a solid rotating object is equal to the product of its rotational inertia and the magnitude of its angular acceleration vector.

(d) The magnitude of the torque vector on a solid rotating object is equal to its rotational inertia divided by the magnitude of its angular acceleration vector.

(e) None of the above

60. Suppose three inductors are connected in parallel. Two of them are rated at 68 mH and one of them is rated at 34 mH. What is the net inductance of this combination, assuming there is no mutual inductance among the components?

(a) 170 mH.

(b) 57 mH.

(c) 54 mH.

(d) 17 mH.

(e) More information is needed to answer this.

61. Neglecting the gravitational effects of the sun and the moon, and assuming the earth is a perfect sphere with uniform density throughout its interior, an object in a perfectly circular orbit around the earth would

(a) continuously gain potential energy at a constant rate.

(b) continuously lose potential energy at a constant rate.

(c) alternately gain and lose potential energy.

(d) always have the same non-zero potential energy.

(e) always have zero potential energy.

62. Suppose a thin, straight wire in free space carries a constant, direct electric current, producing a magnetic flux density of 25.0 G at a point 100 mm from the wire. What is the flux density at a point 500 mm from the wire?

(a) 1.00 G.

(b) 5.00 G.

(c) 11.2 G.

(d) 25.0 G.

(e) 125 G.

63. Refer to Fig. Exam-6. The complex impedance $R + jX$ shown here consists of

    (a) pure resistance.

    (b) pure inductive reactance.

    (c) pure capacitive reactance.

    (d) resistance and inductive reactance.

    (e) resistance and capacitive reactance.

64. The absolute value of the complex impedance shown in Fig. Exam-6 is approximately

    (a) 49.

    (b) 67.

    (c) $49 + j67$.

    (d) 116.

    (e) None of the above

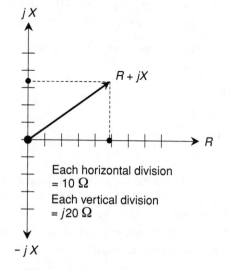

Figure Exam-6 Illustration for Exam Questions 63 and 64.

65. In free space, the product of the frequency of a photon in hertz and its wavelength in meters is equal to

    (a) its period in seconds.

    (b) the speed of light in meters per second.

    (c) Planck's constant in joule-seconds.

    (d) its energy in joules.

    (e) its momentum in kilogram meters per second.

66. When a deuterium (D) nuclide and a tritium (T) nuclide combine in a D-T fusion reaction, the result is a nucleus of He-4 along with energy and

    (a) an emitted neutron.

    (b) an emitted proton.

    (c) an emitted positron.

    (d) an emitted alpha particle.

    (e) an emitted deuteron.

67. When an atomic nucleus undergoes a single gamma-decay event,

    (a) its atomic number and atomic weight both decrease by 1.

    (b) its atomic number stays the same, but its atomic weight decreases by 1.

    (c) its atomic number and atomic weight both stay the same.

    (d) its atomic number decreases by 1, but its atomic weight stays the same.

    (e) its atomic number decreases by 1, and its atomic weight decreases by 2.

68. The net force on an object is equal to the instantaneous rate of change of its

    (a) displacement.

    (b) velocity.

    (c) acceleration.

    (d) momentum.

    (e) mass.

69. Figure Exam-7 shows two hypothetical planets, called No. 1 and No. 2, in circular orbits around a central star. The orbital radii are r_1 and r_2, and the orbital periods are T_1 and T_2 as shown. If the radii are expressed in kilometers and the periods are expressed in seconds, which of the following formulas applies to this situation?

    (a) $T_2 / T_1 = r_2 / r_1$.

    (b) $(T_2 / T_1) = (r_2 / r_1)^2$.

    (c) $(T_2 / T_1)^2 = (r_2 / r_1)$.

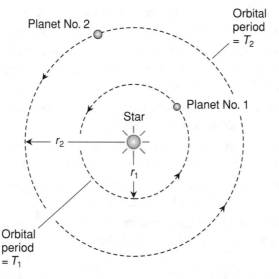

Figure Exam-7 IIllustration for Exam Question 69.

    (d) $(T_2 / T_1)^2 = (r_2 / r_1)^3$.

    (e) We can't state a relation between the orbital radii and the orbital periods unless we also know the ratio of the masses of the planets.

70. When an atomic nucleus undergoes a single beta-decay event,

    (a) its atomic number and atomic weight both stay the same.

    (b) its atomic number stays the same, but its atomic weight decreases by 1.

    (c) its atomic number stays the same, but its atomic weight increases by 1.

    (d) its atomic number and atomic weight both increase or decrease by 1.

    (e) its atomic number increases or decreases by 1, but its atomic weight stays the same.

71. Suppose you drop a 1.00-kg solid object from a height of 12.3 m so it falls straight down onto a concrete floor. What is its kinetic energy an infinitesimal moment before it hits the floor? Consider the acceleration of gravity to be 9.80 m/s². Neglect the effect of air resistance.

    (a) $1.26 \ \text{kg} \cdot \text{m}^2/\text{s}^2$.

    (b) $60.3 \ \text{kg} \cdot \text{m}^2/\text{s}^2$.

    (c) $121 \ \text{kg} \cdot \text{m}^2/\text{s}^2$.

    (d) $781 \ \text{kg} \cdot \text{m}^2/\text{s}^2$.

    (e) $1480 \ \text{kg} \cdot \text{m}^2/\text{s}^2$.

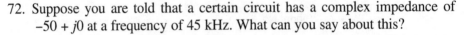

72. Suppose you are told that a certain circuit has a complex impedance of $-50 + j0$ at a frequency of 45 kHz. What can you say about this?

    (a) The person who told you this is mistaken, or else you have misinterpreted what was said. In practice, a circuit cannot have a complex impedance of $-50 + j0$.

    (b) The circuit contains resistance, but no inductive or capacitive reactance, at a frequency of 45 kHz.

    (c) The circuit contains resistance and capacitive reactance, but no inductive reactance, at a frequency of 45 kHz.

    (d) The circuit contains resistance and inductive reactance, but no capacitive reactance, at a frequency of 45 kHz.

    (e) The circuit contains resistance, inductive reactance, and capacitive reactance at a frequency of 45 kHz.

73. Power is an expression of

    (a) energy accumulated over time.

    (b) the rate at which work is done.

    (c) the instantaneous rate of change in acceleration of an object.

    (d) the accumulated force on an accelerating object.

    (e) Any of the above

74. Imagine that you're on the surface of a perfectly spherical planet made up of solid rock having uniform density throughout. The radius of the planet is 4932 km. The acceleration of gravity at the surface is 8.765 m/s². Engineers have dug a shaft 731 km straight down into the planet. When scientists descend to the bottom of the shaft to assemble a laboratory to detect neutrinos from distant galaxies, what is the acceleration of gravity they feel?

    (a) Zero.

    (b) 0.650 m/s².

    (c) 1.30 m/s².

    (d) 7.47 m/s².

    (e) 8.12 m/s².

75. In any medium of uniform constitution and density, the force of repulsion between two tiny spherical objects X and Y that are a constant distance from each other, and that carry positive electric charge quantities q_x and q_y respectively, is

    (a) inversely proportional to $q_x + q_y$.

    (b) directly proportional to $q_x + q_y$.

(c) inversely proportional to $q_x q_y$.

(d) directly proportional to $q_x q_y$.

(e) directly proportional to the absolute value of $q_x - q_y$.

76. In power-of-10 expressions for physical units, the prefix multiplier *nano-* means

(a) one-thousandth (10^{-3}).

(b) one-millionth (10^{-6}).

(c) one-thousand-millionth (10^{-9}).

(d) one-trillionth (10^{-12}).

(e) one-quadrillionth (10^{-15}).

77. The absolute value of a complex number is always

(a) a non-negative real number.

(b) an integer.

(c) a non-negative imaginary number.

(d) between and including 0 and 2π.

(e) smaller than its real-number component.

78. Suppose you push a 4.27-kg block for a distance of 8.80 m up a friction-less, straight inclined plane tilted at an angle of 30° relative to the horizontal. How much work is expended if the apparatus is located on a planet where the acceleration of gravity is 6.45 m/s²?

(a) 121 N · m.

(b) 210 N · m.

(c) 242 N · m

(d) 2.13 × 103 N · m.

(e) More information is needed to answer this.

79. In free space, the ratio of the energy of a photon in joules to its frequency in hertz is equal to

(a) its period in seconds.

(b) the speed of light in meters per second.

(c) Planck's constant in joule-seconds.

(d) its energy in joules.

(e) its momentum in kilogram meters per second.

80. Fill in the blank to make the following sentence true: "The number of joules it takes to raise or lower the temperature of 6.02×10^{23} atoms of a substance by 1 K when there is no change of state is called the _____ of that substance."

    (a) atomic heat coefficient

    (b) thermal energy transfer quantity

    (c) molar heat capacity

    (d) specific thermal entropy

    (e) heat of fusion

81. Miles per hour squared could, in theory, be used as a unit of

    (a) angular acceleration.

    (b) linear speed.

    (c) time per unit area.

    (d) linear velocity.

    (e) gravitational field strength.

82. As viewed from high above the north geographic pole, the earth rotates counterclockwise at a constant rate. Remember from your astronomy courses that:

    - The north celestial pole is a distant imaginary point in the heavens over the north geographic pole

    - The south celestial pole is a distant imaginary point in the heavens over the south geographic pole

    - The celestial equator is a huge imaginary circle in the heavens over the geographic equator

    Based on this information, we can say that the angular momentum vector of the earth

    (a) has constant magnitude and points away from the center of the earth toward the north celestial pole.

    (b) has variable magnitude and points in a direction away from the center of the earth and alternating between the north and south celestial poles, completing one cycle with every complete rotation of the planet.

    (c) has constant magnitude and points away from the center of the earth toward the south celestial pole.

    (d) has constant magnitude and points inward from the celestial equator toward the center of the earth, rotating through a full circle with every complete rotation of the planet.

    (e) cannot be defined because the planet is spherical, not disk-shaped.

83. What is the electric flux density through a sphere having a diameter of 5.0 m with a point charge of +1.0 C at its center?

    (a) 0.40 C/m.

    (b) 0.013 C/m².

    (c) 0.015 C/m³.

    (d) 0.064 m/C.

    (e) More information is needed to answer this.

84. Figure Exam-8 is a hypothetical plot of the relative particle-emission frequency from a sample of material undergoing beta decay. What, if anything, is theoretically wrong with this plot?

    (a) Beta decay always occurs within days or weeks, not millions of years.

    (b) The rate of emission should be constant, not decreasing with time.

    (c) The plot should not be a straight line, but a curve whose derivative is positive and approaches zero as time passes.

    (d) The plot should not be a straight line, but a curve whose derivative is negative and approaches zero as time passes.

    (e) Nothing is wrong here. This rendition of beta decay is perfectly plausible.

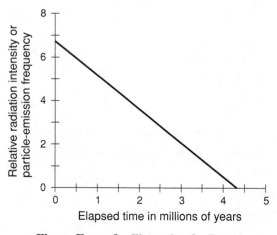

Figure Exam-8   Illustration for Exam Question 84.

85. All of the following are common processes that maintain the plasma in a tokamak at a high temperature, *except*

    (a) the transmission of gamma rays into the plasma.

    (b) ohmic heating caused by electric current in the plasma.

    (c) self-heating that results from the ongoing fusion reaction.

    (d) the firing of neutral atoms into the plasma at high speed.

    (e) the transmission of radio waves into the plasma.

86. The Bohr radius is a physical constant defined as

    (a) the radius of a proton.

    (b) the radius of an alpha particle.

    (c) the radius of the highest-energy electron shell in a uranium atom.

    (d) the radius of the lowest-energy electron shell in a hydrogen atom.

    (e) the average distance between the atoms of pure hydrogen gas at standard temperature and pressure.

87. Here are the formulas for inductive reactance jX_L and capacitive reactance jX_C in terms of inductance L, capacitance C, and frequency f:

$$jX_L = j2\pi fL$$
$$jX_C = -j(2\pi fC)^{-1}$$

where L is in microhenrys, C is in microfarads, and f is in megahertz. Given these formulas, it is possible to derive a formula for the frequency f of a series-resonant circuit in terms of L and C. What is this formula? Remember that at the resonant frequency, the net reactance is zero.

    (a) $f = (2\pi LC)^{-1}$.

    (b) $f = 2\pi LC$.

    (c) $f = [(2\pi) (LC)^{1/2}]^{-1}$.

    (d) $f = 2\pi (LC)^{1/2}$.

    (e) $f = (2\pi LC)^{-1/2}$.

88. In Fig. Exam-9, which, if any, of the vectors shown represents a conductance in parallel with an inductive susceptance?

    (a) Vector **A**.

    (b) Vector **B**.

    (c) Vector **C**.

    (d) Vector **D**.

    (e) None of them.

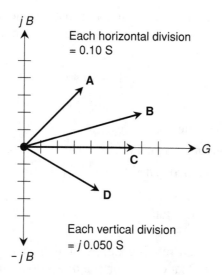

Each horizontal division
= 0.10 S

Each vertical division
= j 0.050 S

Figure Exam-9 Illustration for Exam
Questions 88 and 89.

89. In Fig. Exam-9, which, if any, of the vectors shown represents a pure, finite, non-zero resistance?

(a) Vector **A**.

(b) Vector **B**.

(c) Vector **C**.

(d) Vector **D**.

(e) None of them.

90. Let P be the pressure of an ideal gas having absolute temperature T, confined to an enclosure having volume V. Let N represent the number of atoms or molecules of gas in the enclosure. Then:

$$P = RNT / V$$

where P is in pascals, T is in kelvins, V is in meters cubed, N is in moles, and R is a constant expressed in

(a) newton meters per second.

(b) kilogram meters per second squared.

(c) joules per mole per kelvin.

(d) meters cubed per second squared.

(e) newton kilograms per meter.

91. For a rotating solid object, suppose that $\boldsymbol{\tau}$ is the net torque vector, $\mathbf{0}$ is the zero vector, $\mathbf{L}$ is the angular momentum vector, and $\mathbf{k}$ is a constant vector. Then:

$$(\boldsymbol{\tau} = \mathbf{0}) \Leftrightarrow (\mathbf{L} = \mathbf{k})$$

This expresses the law of conservation of

(a) rotational potential energy.

(b) rotational mechanical energy.

(c) rotational kinetic energy.

(d) torque.

(e) angular momentum.

92. Imagine two non-rotating lumps of light clay-like stuff, both of mass 1.50 kg. They float on the surface of a smooth pond. There is no friction between the objects and the water, no current or other water turbulence in the pond, and no wind to affect the objects or make waves on the pond surface. Both lumps move horizontally in straight lines on a collision course at 4.00 m/s. Before they hit, one lump travels due north and the other lump travels due east. When they collide, they stick together. What is the magnitude of the momentum vector of the composite lump after the collision?

(a) 6.00 kg · m/s.

(b) 8.49 kg · m/s.

(c) 9.00 kg · m/s.

(d) 12.0 kg · m/s.

(e) More information is necessary to answer this.

93. Figure Exam-10 shows two complex admittance vectors, $\mathbf{Y}_1$ and $\mathbf{Y}_2$, plotted in the GB plane. The dashed lines connecting the terminating points of the vectors with the axes indicate the component conductances and susceptances. This plot tells us that a parallel combination of components having these admittances at a certain frequency f will exhibit

(a) finite, non-zero conductance and inductance at f.

(b) finite, non-zero conductance and capacitance at f.

(c) parallel resonance at f.

(d) the properties of a short circuit at f.

(e) the properties of an open circuit at f.

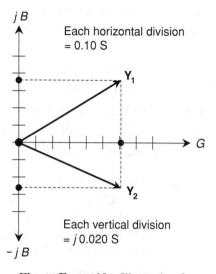

jB

Each horizontal division
= 0.10 S

Y_1

G

Y_2

Each vertical division
= j 0.020 S

$-jB$

Figure Exam-10 Illustration for
Exam Question 93.

94. The Schwarzchild radius of an object depends on
    (a) its mass and radius.
    (b) its radius only.
    (c) its mass only.
    (d) neither its mass nor its radius.
    (e) the intensity of the gravitational field at the surface.

95. Imagine two parallel-wire transmission lines P and Q that are both
    terminated in ideal loads having complex impedances of $R_p + j0$ and $R_q + j0$.
    Also suppose that the input (non-load) ends of both lines are connected
    to the same source of constant AC voltage, and that both lines have ab-
    solutely no power loss. If the characteristic impedance of line P is exactly
    one-quarter of the characteristic impedance of line Q, then how do the cur-
    rents I_p and I_q through the two ideal loads $R_p + j0$ and $R_q + j0$ compare?
    (a) $I_p = 16 I_q$.
    (b) $I_p = 4 I_q$.
    (c) $I_p = 2 I_q$.
    (d) $I_p = I_q$.
    (e) None of the above

96. If all the atoms or molecules in a sample of gas have the same mass and the temperature is uniform throughout the sample, then the rms speed of the particles is

    (a) directly proportional to the square root of the absolute temperature.

    (b) directly proportional to the absolute temperature.

    (c) directly proportional to the square of the absolute temperature.

    (d) inversely proportional to the absolute temperature.

    (e) independent of the absolute temperature.

97. Imagine an electrically non-conducting sphere of variable radius with a single point electric charge of exactly 1 C at its center. Suppose the interior of the sphere is a vacuum, and the sphere is suspended in a vacuum with no other objects nearby. How does the total electric flux quantity Φ passing through the sphere vary as the radius r of the sphere varies?

    (a) The value of Φ is directly proportional to r.

    (b) The value of Φ is inversely proportional to r.

    (c) The value of Φ is directly proportional to r^2.

    (d) The value of Φ is inversely proportional to r^2.

    (e) The value of Φ does not depend on r.

98. Suppose a liquid has a specific heat of 0.75 J/kg/K. If 1.4 J of thermal energy is transferred to a 2.0-kg sample of this substance, and assuming it remains liquid during the entire process, which of the following will happen?

    (a) The temperature of the sample will rise by 3.7 K.

    (b) The temperature of the sample will rise by 2.1 K.

    (c) The temperature of the sample will rise by 0.93 K.

    (d) The temperature of the sample will rise by 0.53 K.

    (e) The temperature of the sample will not change.

99. Suppose you drop a ball from a great height. Let $\mathbf{q}(t)$ be the function that defines the instantaneous downward displacement vector of the ball, relative to its starting point, versus time t, where $t = 0$ at the instant you let the ball go. Let $\mathbf{v}(t)$ be the function that defines the instantaneous velocity vector of the with respect to time, with $\mathbf{v} = \mathbf{0}$ at $t = 0$. Consider the acceleration of gravity to be 10 m/s^2, and assume there is no air resistance. Which of the following equations is true until the ball hits the surface?

    (a) $\mathbf{v}(t) = 10 \, \mathbf{q}(t) \, / \, dt$.

    (b) $\mathbf{v}(t) = 0.10 \, \mathbf{q}(t) \, / \, dt$.

(c) $\mathbf{v}(t) = d\mathbf{q}(t) / dt$.

(d) $\mathbf{v}(t) = 10 \, d\mathbf{q}(t) / dt$.

(e) $\mathbf{v}(t) = 0.10 \, d\mathbf{q}(t) / dt$.

100. Let p represent the potential energy that an object possesses. Let n represent its kinetic energy, and let m represent its mechanical energy. Which of the following formulas holds true according to the work-energy theorem, provided that all three forms of energy are expressed in equivalent units?

(a) $p = m - n$.

(b) $n = m + p$.

(c) $n = p - m$.

(d) $m = n^2 - p$.

(e) $m = p^2 + n$.

Answers to Quiz and Exam Questions

Chapter 1

1. b	2. c	3. d	4. c	5. a
6. a	7. c	8. d	9. b	10. b

Chapter 2

1. a	2. c	3. c	4. c	5. b
6. b	7. d	8. c	9. a	10. c

Chapter 3

1. d	2. d	3. d	4. b	5. d
6. c	7. a	8. c	9. a	10. b

Chapter 4

1. d	2. a	3. b	4. a	5. c
6. b	7. d	8. b	9. d	10. a

Chapter 5

1. a	2. d	3. d	4. c	5. c
6. d	7. a	8. b	9. c	10. d

Chapter 6

1. b	2. a	3. b	4. a	5. d
6. a	7. c	8. d	9. d	10. c

Chapter 7

1. b	2. a	3. d	4. c	5. a
6. b	7. a	8. b	9. c	10. b

Chapter 8

1. c	2. d	3. c	4. a	5. b
6. c	7. d	8. b	9. b	10. b

Chapter 9

1. a	2. d	3. a	4. b	5. c
6. c	7. c	8. c	9. a	10. d

Final Exam

1. a	2. c	3. c	4. e	5. e
6. b	7. b	8. d	9. d	10. a
11. d	12. e	13. b	14. c	15. c
16. e	17. c	18. e	19. a	20. c
21. d	22. a	23. e	24. c	25. a
26. a	27. a	28. a	29. b	30. e
31. a	32. c	33. d	34. b	35. e
36. e	37. d	38. c	39. e	40. b
41. b	42. c	43. b	44. d	45. d
46. a	47. a	48. c	49. e	50. d
51. d	52. d	53. b	54. b	55. a
56. e	57. d	58. c	59. c	60. d
61. d	62. b	63. d	64. e	65. b
66. a	67. c	68. d	69. d	70. e
71. c	72. a	73. b	74. d	75. d
76. c	77. a	78. a	79. c	80. c
81. e	82. a	83. b	84. d	85. a
86. d	87. c	88. d	89. c	90. c
91. e	92. b	93. b	94. c	95. b
96. a	97. e	98. c	99. c	100. a

Suggested Additional Reading and Reference

Books

- Gautreau, R. and Savin, W., *Schaum's Outline of Modern Physics*. New York, NY: McGraw-Hill, 1999.
- Giancoli, D., *Physics: Principles with Applications, 6th edition*. San Francisco, CA: Benjamin Cummings, 2004.
- Gibilisco, S., *Astronomy Demystified*. New York, NY: McGraw-Hill, 2003.
- Gibilisco, S., *Physics Demystified*. New York, NY: McGraw-Hill, 2002.
- Gibilisco, S., *Teach Yourself Electricity and Electronics, 4th edition*. New York, NY: McGraw-Hill, 2006.
- Gibilisco, S., *Technical Math Demystified*. New York, NY: McGraw-Hill, 2006.

- Halpern, A., *3000 Solved Problems in Physics*. New York, NY: McGraw-Hill Professional Publishing, 1988.
- Krantz, S., *Calculus Demystified*. New York, NY: McGraw-Hill, 2003.
- McMahon, D., *Quantum Mechanics Demystified*. New York, NY: McGraw-Hill, 2005.
- McMahon, D. and Alsing, P., *Relativity Demystified*. New York, NY: McGraw-Hill, 2005.

Web Sites

- *Encyclopedia Britannica Online:* www.britannica.com
- *Eric's Treasure Troves of Science:* www.treasure-troves.com
- *National Institute of Standards and Technology Physics Laboratory:* www. physics. nist.gov
- *PhysicsWeb:* physicsweb.org

INDEX